Theory and Applications of Transport in Porous Media

Volume 28

More information about this series at http://www.springer.com/series/6612

Hui-Hai Liu

Fluid Flow in the Subsurface

History, Generalization and Applications of Physical Laws

Springer

Hui-Hai Liu
Aramco Research Center
Aramco Services Company
Houston, TX
USA

ISSN 0924-6118 ISSN 2213-6940 (electronic)
Theory and Applications of Transport in Porous Media
ISBN 978-3-319-43448-3 ISBN 978-3-319-43449-0 (eBook)
DOI 10.1007/978-3-319-43449-0

Library of Congress Control Number: 2016947035

Printed on acid-free paper

This Springer imprint is published by Springer Nature
The registered company is Springer International Publishing AG Switzerland

Preface

This research monograph presents a systematic endeavor to generalize fundamental physical laws related to subsurface fluid flow that are important for a number of contemporary applications, such as recovery of subsurface energy resources, geological disposal of high-level nuclear wastes, CO_2 geological sequestration, and groundwater contamination in the vadose zone. The history of discovering these laws is also briefly presented within the context of discussing the ranges in which they are valid. This relatively new endeavor should be of interest to engineers, researchers, and students in the areas of reservoir engineering, hydrogeology, soil physics, and rock mechanics.

Darcy's law is the fundamental law of subsurface fluid flow. For low-permeability porous media, however, Darcy's law does not always hold because of the strong fluid–solid interaction. While this issue has been investigated by a number of researchers, Chap. 1 presents a new phenomenological relationship between water flux and hydraulic gradient, or a generalized Darcy's law. The traditional form of Darcy's law and two other generalizations for low-permeability media, proposed by other researchers, are shown to be special cases of our generalized Darcy's law. The implications and applications of this "non-Darcian" flow behavior are also discussed.

Edgar Buckingham independently discovered the relationship between water flux and hydraulic gradient for unsaturated (or multiphase) flow in porous media that has often been called Darcy's law in the literature. In this book, I call this relationship the Darcy-Buckingham law because in my view Buckingham's contribution to the theory of subsurface multiphase flow has been historically underestimated; multiphase flow is more complex than the single-phase flow that Darcy's law was developed for. The Darcy-Buckingham law, however, is only valid under the condition of local equilibrium. This condition does not hold in many cases, although the Darcy-Buckingham law has been used there because of the lack of alternatives. In Chap. 2, I introduce an optimality principle that unsaturated water flow patterns are self-organized in such a way that overall water flow resistance is minimal. Based on this principle, a generalized version of the Darcy-Buckingham law is derived in which unsaturated hydraulic conductivity is not only a function of water saturation (or capillary pressure) assumed in the Darcy-Buckingham law, but also a function of water flux. The consistency between our theoretical results and observations is demonstrated.

It is well known that subsurface fluid flow is coupled with mechanical deformation of subsurface media where fluid flow occurs; in many cases, this coupling can play a dominant role. Hooke's law is the most fundamental law governing elastic deformation of solids. Natural rocks, however, have unique features compared with other solids, one of which is the existence of small-scale deformation heterogeneities (such as microcracks). Acknowledgement of these unique features is important, because they justify that rock mechanics exists as a stand-alone discipline, rather than an adjunct of general solid mechanics. This also explains why elastic mechanical deformation of a natural rock does not follow exactly the traditional Hooke's law; the related mechanical properties, unlike those assumed by Hooke's law, are not constant under certain conditions. To better consider the impact of natural heterogeneity, Chap. 3 introduces the two-part Hooke model that was developed by dividing a natural rock into hard and soft parts. Remarkable consistency between the model and observations from different sources, for both rock matrix and fractures, has been achieved. The usefulness of the model in dealing with engineering problems is also demonstrated. Note that although a number of researchers have touched on the same issue by establishing empirical relations between stress and rock mechanical properties, the two-part Hooke model takes a much bolder and more systematic approach and is also more effective for practical applications.

Non-equilibrium thermodynamics (which is closely related to optimality principles) is the foundation for dealing with highly nonlinear problems. This branch of science, however, has not been well established yet. For example, contradictory optimality principles exist in the literature. As an applied scientist and engineer with a primary interest in applying basic scientific principles to my research areas, I initially tried to avoid, but eventually got into the study of the nonequilibrium thermodynamics because it is the true starting point to investigate subsurface multiphase flow and other related processes. Chapter 4 presents a new thermodynamics hypothesis that tries to answer under what conditions the optimality principles should apply and which of the simultaneously occurring physical processes, if not all, is subject to the optimization. This hypothesis seems to be able to reconcile different optimality principles proposed in several different areas.

The generalization of well-known fundamental physical laws is indeed an ambitious and a highly risky endeavor. It, however, was not motivated by academic interest, but by the needs of practical applications. As a modeler who is fully aware of and enjoys the increasing powerfulness of available computational capabilities, I am more and more convinced that the lack of appropriate physical laws at scales of practical interest is the weakest link in improving our modeling capability (especially the capability for prediction). For example, no matter how powerful computers are, we simply cannot use them to predict the observed non-Darcian flow with the traditional Darcy's law. Much more work is needed to substantiate our physical foundations for accurately modeling subsurface processes.

The generalization of the physical laws uses different approaches in this book, ranging from purely phenomenological ones to theoretical derivations based on newly introduced principles. But all generalized laws have the relatively simple mathematical form and a small number of parameters. This largely reflects my own research philosophy as an engineer: a good model or theory should be able to adequately capture the essence of physics and, at the same time, has a relatively simple form or structure. This happens to be consistent with the point of view of Nobel Laureate Richard Feynman, who in his celebrated book "*Characters of Physical Laws*", concluded that the mathematical forms of well-known physical laws are always simple, although they describe very complex phenomena.

I did initially have a reservation to publish a book dealing with the physical laws that are at the heart of several research areas. I would feel much more comfortable doing so after the related work becomes more mature. At the same time, I also feel the urgency to get the message out that we do need to revisit the fundamental laws that have generally been viewed as sacrosanct doctrines by many people. Thus, I view this book as a messenger or a starting point for this revisiting. I would also like to make it clear that the focus of this book is on the work mainly done by me and my collaborators. Although I try to briefly cover the related work of others as well, the citation of their work is by no means intended to be exhaustive.

Houston, TX, USA Hui-Hai Liu

Acknowledgments

I thank Prof. Fred J. Molz at Clemson University, South Carolina, for introducing me to the area of nonlinear dynamics that is, directly or indirectly, related to several topics discussed in this book. Fred is a top thinker and well-recognized scientific leader in the hydrogeology community. Since I knew him when I was a graduate student, he has been a great friend of mine. Fred always focuses on solving important problems in his field, emphasizes both the problem solving and physical understanding, and has an admirable capability to link fundamental physics and practical applications. He never stops learning new things and has worked very hard to keep bringing fresh ideas/concepts into his field. Readers can easily recognize his significant impacts on my research philosophy from this book.

I have been blessed in my career by being exposed to projects of national and international importance and by having some world-class scientists/engineers as colleagues in both Lawrence Berkeley National Laboratory and Aramco Research Center. Many materials here result from dealing with technical issues of these projects and from collaborations with colleagues whose names can be found from the author lists of the cited papers or reports (in the reference sections) that I coauthored. I appreciate their company in my journey to revisiting the relevant physical laws that could have been very lonely otherwise. Especially, I like to thank some of my postdoc researchers and research students working with me when I was in Lawrence Berkeley National Laboratory; some contents of this book are directly connected to their work and/or stimulated by discussions with them. They are Drs. Feng Sheng, Lianchong Li, Yu Zhao, Jianping Zuo, Marco Bianchi, Mingyao Wei, Run Hu, Jiangtao Zheng, Delphine Roubinet, and Emma Engström. Dr. Run Hu also helped with preparing the initial version of this book. I have had a chance to work with many young and talent Chinese colleagues, because one of my responsibilities at Lawrence Berkeley National Laboratory was to help to establish connections between the Earth Sciences Division over there and Chinese research organizations in the relevant areas.

The draft version of this book was reviewed by Prof. Fred J. Molz from Clemson University (Chaps. 2 and 4), Prof. YuShu Wu from Colorado School of Mines

(Chaps. 1 and 5), and Dr. Yanhui Han from Aramco Research Center-Houston (Chap. 3). Mr. Jilin Zhang from Aramco Research Center-Houston reviewed the initial version of book and helped check its final version sent to the publisher. Their constructive comments considerably improved the readability of the book. Ms. Ebony N. Fondren from Aramco Research Center-Houston took care of the paperwork related to copyright issues for previously published illustrations used in this book. I also thank the management of the Aramco Research Center and Saudi Aramco, especially Drs. Ashraf M. Al-Tahini (Director of Aramco Research Centers) and Daniel T. Georgi (Lead of Reservoir Engineering Technology Team from Aramco Research Center—Houston), for supporting the publication of this book.

I am indebted to Ms. Nishanthi Venkatesan, Ms. Petra van Steenbergen, and Ms. Hermine Vloemans from Springer for their guidance, patience, and help during writing and publishing this book. They made the whole process much more pleasant than it might have been.

I am grateful to Elsevier, Swiss Geological Survey at Swisstopo, and Nagra for their permissions to use previously published illustrations in this book.

Finally, I very much appreciate the support and encouragement from my wife Wei Huang and my son Neil Liu while writing this book. I initially thought that writing a book was something that I should do after retirement. The other consideration is that the related work may become more mature at that time. Wei and Neil were able to make me change my mind based on the argument that writing a book to summarize previous research work does not necessarily slow down the current research activities; instead it provides an opportunity to rethink some important issues. It turns out that they are right!

Contents

Chapter 1
Generalization of Darcy's Law: Non-Darcian Liquid Flow in Low-Permeability Media

Abstract Darcy's Law was discovered by Henry Darcy (1803–1858) based on experimental observations of steady-state water flow through sand columns. It states that water flux in saturated porous media is linearly proportional to hydraulic gradient. For low-permeability porous media, however, Darcy's law is not adequate because of the strong fluid–solid interaction that results in non-linear flux-gradient relationships. This chapter presents a new phenomenological relationship between water flux and hydraulic gradient, or a generalized Darcy's law. The traditional form of Darcy's law and two other generalizations for low-permeability media, proposed by other researchers, are shown to be special cases of the generalization. The consistency between the generalization and experimental observations from different sources is demonstrated. The generalized Darcy's law and its variations are used to attack several key technical issues facing the geoscience community, including the relative importance of diffusion in the excavation damaged zone for a shale repository of high-level nuclear waste, the accurate measurement of relative permeability for multiphase flow in a low-permeability porous medium, non-Darcian flow behavior during imbibition of fracturing fluids into a shale gas reservoir, and formation of the pressure seal in shale formations.

Darcy's law is the fundamental law for modeling subsurface fluid flow processes. This chapter briefly reviews Darcy's life and his law and presents a generalized Darcy's law for low-permeability media in which traditional Darcy's law does not hold for describing liquid flow. Applications of the generalization to several practical problems are also discussed. The discussions in this chapter are mainly based on materials from Liu et al. (2012, 2014), Liu and Birkholzer (2013) and Liu (2014).

H.-H. Liu, *Fluid Flow in the Subsurface*, Theory and Applications of Transport in Porous Media 28, DOI 10.1007/978-3-319-43449-0_1

1.1 Henry Darcy and His Law for Subsurface Fluid Flow

The well-known Darcy's Law was discovered by Henry Darcy (1803–1858). It states that water flux in saturated porous media is linearly proportional to hydraulic gradient. Darcy's Law forms the quantitative basis of many science and engineering disciplines including hydrogeology, soil science, and reservoir engineering. Because of the unparalleled status of Darcy's law in the related areas, there are a number of excellent publications in the literature regarding Darcy's life and his law (e.g., Freeze 1994; Brown 2002; Simmons 2008). The materials presented in this section are based on these publications.

Henry Darcy (Fig. 1.1) was a distinguished engineer, scientist and citizen (Brown 2002). He was born on 10 June 1803 in Dijon, France, and entered L'Ecole Polytechnique, Paris, in 1821, where he started his science and engineering training. At that time Jean-Baptiste Joseph Fourier (1768–1830), who discovered Fourier Law for heat transfer, held a Chair position at L'Ecole Polytechnique. Thus, Simmons (2008) speculated that "it is therefore possible that Fourier taught Darcy his heat law and that the earliest seeds of Darcy's Law may have been planted at this point". Darcy was admitted to L'Ecole des Ponts et Chaussèes (School of Bridges and Roads), Paris, in 1823 when he was 20 years old. The school was closely associated with Le Corps Des Ponts et Chaussèes that had a mission to support the construction of infrastructure throughout France and was regarded an elite fraternity of engineers that had influential status in mid-nineteenth century in France (Freeze 1994; Simmons 2008). The school was created to train students to

Fig. 1.1 Henry Darcy (1803–1858) (https://en.wikipedia.org/wiki/Henry_Darcy)

be engineers in the Corps. Based on Darcy's class ranking in both L'Ecole Polytechnique and L'Ecole des Ponts et Chaussées, Brown (2002) noted that Darcy was a good, but not the best student in his classes.

In 1826, Darcy, at the age of 23, graduated from the School of Bridges and Roads and started a remarkable scientific and engineering career in the Corps. He spent almost all of his working life in his home city, Dijon. During 1827–1834, he performed feasibility studies of, and developed a plan for, the Dijon public water supply project. The plan was approved in 1835 and the project began in 1839 and was widely regarded one of the best European water supply systems at that time (Simmons 2008). Because of the success of the project, Darcy received a number of honors. He was awarded the Legion of Honor by King Louis Phillip in 1842. He also accepted a gold medal from the Municipal Council and a laurel wreath from the workmen when the project was completed in 1844, but he waived all the fees. As Philip (1995) put it, "Darcy, with great vision and skill, designed and built a pure water supply system for Dijon, in place of previous squalor and filth. Dijon became a model for the rest of Europe. Darcy selfless waived fees due to him from the town, corresponding to about $1.5 million today. Medals were struck recognizing his skill and a monument celebrates his great work".

Darcy's very success lay in his downfall (Philip 1995; Simmons 2008). During the last decade of his life (1848–1858), he suffered political persecution and his health deteriorated. Fortunately, he was able to focus on research activities very productively in this relatively short period of his life and made major scientific discoveries, including Darcy's Law. The period was called Darcy scientific legacy by Simmons (2008). As an engineer, Darcy's scientific contributions were clearly motivated by a deep desire to solve practical and relevant engineering problems that he had encountered. Darcy's other contributions than Darcy's law and the related historical backgrounds can be found in details in Simmons (2008).

In 1856, Darcy published his most famous report on the construction of municipal water supply of Dijon, "Les Fontaines Publiques de la Ville de Dijon", or "The public fountains of the city of Dijon" in English. The report was about 680 pages long and contained 28 plates of figures. Part 2 of Note D of the report, with a subtitle of "Determination of the laws of water flow through sand", contains the results of his sand column experiments conducted with the set-up shown in Fig. 1.2. The motivation for the experiments was to investigate water flow through a sand filter. At that time, the water filtration method was a common practice to improve water clarity and, as a result, engineers were starting to think about the water flow behavior through filters (Simmons 2008). In the experiments, Darcy and his assistant applied the water on the top of the sand column (under the saturated condition), took measurements of discharge (flow) rate at the bottom, and at the same time monitored the water hydraulic heads at two points near the top and bottom (Fig. 1.2). Details of the experimental procedures and the observations are discussed in Freeze (1994), Brown (2002) and Simmons (2008). Using the obtained

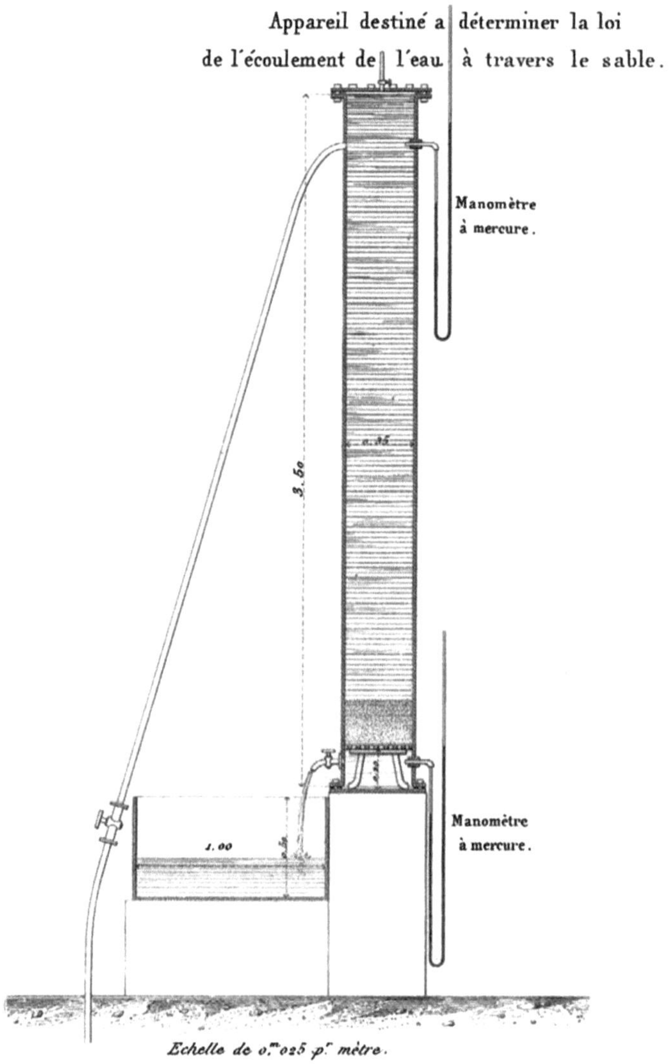

Fig. 1.2 Darcy sand column apparatus (Darcy 1856)

data, Darcy discovered an empirical relationship (Fig. 1.3), now called Darcy's Law, that the water flow rate is proportional to the cross-sectional area of the sand column and hydraulic gradient. This statement of Darcy's law is equivalent to the second sentence of this section.

Darcy passed away in 1858, two years after he published the fountain report. Although the Darcy's law was buried in the depth of the report, Simmons (2008) suggests that "Darcy understood that his discovery was new and significant". Supporting evidence includes the fact that Darcy dedicated almost half of the length

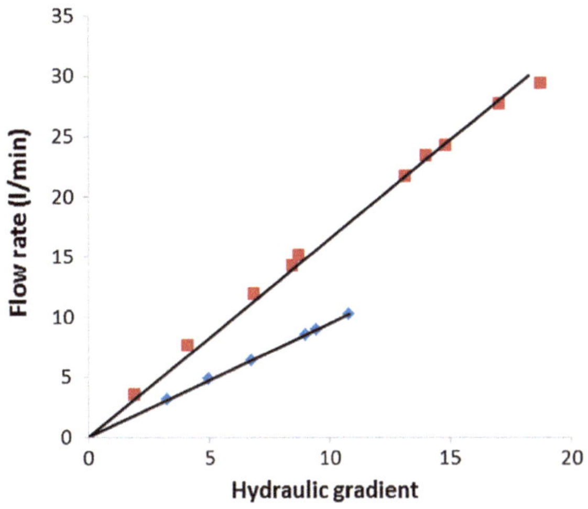

Fig. 1.3 Darcy's sand-column experimental results (the red and blue *squares*) (Darcy 1856). The *solid lines* are best linear fitting curves

of his preface to his report to a discussion on Darcy's Law. He also mentioned in the preface: "I have not seen the documents that are included in Note D collected in any special book. In particular, to my knowledge at least, no one has experimentally demonstrated the law of water flow through sand".

One may wonder why it was an "engineer" (Henry Darcy), not more theoretically oriented "scientists", who discovered such an important scientific (physical) law. It is a reasonable question to ask, because fluid mechanics and hydrodynamics theories had been well established at that time (Brown 2002). There might be many reasons for that. One speculation is that, when applying these theories, people often got stuck with details, such as the "messy" solid–water boundaries in a porous medium, and failed to see the key picture that pore spaces can be conceptually linked to the water flow in a capillary tube. Had they been able to, it would have been straightforward to deduce Darcy's law from the existing fluid mechanics theories, such as Poiseuille equation that gives pressure drop in a fluid flowing through a cylindrical pipe. This issue will be discussed further in Sect. 1.2. Thus, Darcy's success in discovering Darcy's law is largely because he was able to see the simplicity out of complexity. There are many famous quotes regarding simplicity and complexity in the literature. The quote that the author of this book likes the most is the following one from Steve Jobs, the late co-founder and CEO of Apple Inc. "That's been one of my mantras—focus and simplicity. Simple can be harder than complex; you have to work hard to get your thinking clean to make it simple." Although Jobs's statements are actually about technology invention and business, they can be equally applied to scientific discoveries as well.

While Darcy discovered Darcy's law using an empirical approach, he had the conceptual linkage in his mind and just used the test data to confirm his prediction (Simmons 2008). Darcy noted clearly in footnote 4 of Note D in his report, "I had already foreseen this curious result in my research on water flow in conduit pipes of very small diameters, …".

An interesting observation is that Darcy formally discovered his law at the age of 53. We are often told, especially by theoretical physicists, that one likely makes his or her most important scientific contributions at very young ages. Darcy's story suggests that it is not necessarily true at least for applied scientists or engineers who may have much longer career periods for significantly creative activities. This is really encouraging for many people in the engineering fields (including the author of this book) who are not very young now!

1.2 Relationship Between Water Flow Flux and Hydraulic Gradient in a Capillary Tube

The similarity of water flow in porous media to that in a capillary tube, first realized by Darcy (1856), is an important concept for studying flow and transport processes in porous media. Following this line, Dupuit (1857), according to Narasimhan (2005), further conceptualized the pore space in a porous medium as a collection of capillary tubes. This concept is critical for relating hydraulic properties to the pore size distribution for a porous medium and for extending Darcy's law to multiphase flow conditions (Chap. 2). A detailed derivation of a relationship between water flux and hydraulic gradient for a capillary tube with radius R (Fig. 1.4) is given

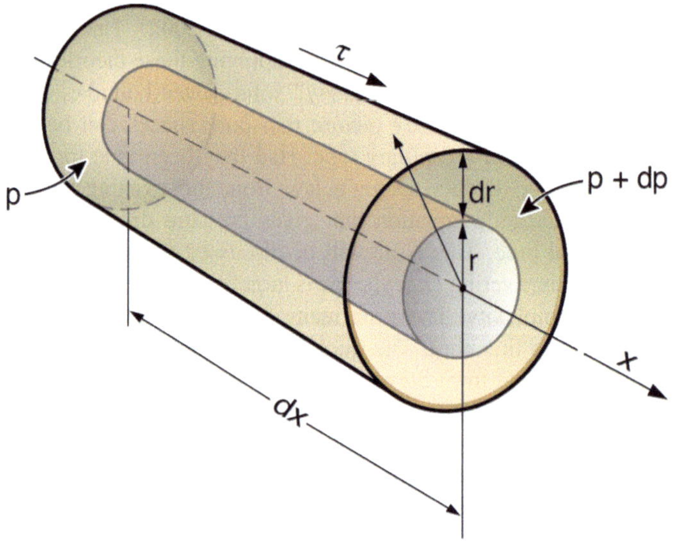

Fig. 1.4 A water element in a horizontal capillary tube with radius R (Liu et al. 2012) (reproduced by permission of Elsevier). The variable r is the radius of a water element within the capillary tube and ranges from zero to R

here; readers familiar with the subject may skip this section. For simplicity, we consider a horizontal capillary tube here, although this relationship could be easily extended to capillary tubes with other orientations.

For laminar water flow in the capillary tube, the shearing stress is given by

$$\tau = -\mu \frac{du}{dr} \tag{1.1}$$

where τ is shearing stress, μ is water viscosity, u is water velocity along the longitudinal direction of the capillary tube, and r is the radial coordinate. For a water surface with radius r and length dx (Fig. 1.4), the total shearing force (F) acting on the cylindrical surface is equal to the shearing stress multiplied by the corresponding surface area, or

$$F = \tau(2\pi r)dx \tag{1.2}$$

Then, the net shearing force acting on a water element (shown in Fig. 1.4) with thickness dr within the capillary tube, dF, is given by differentiating Eq. 1.2 with respect to r, or $dF = 2\pi(dx)d(r\tau)$. For laminar flow, the inertial effect can be ignored because of the low water velocity. In this case, the water element should be subject to zero net force, with the shearing force being balanced by an opposing force (resulting from the pressure difference imposed on the water element) that is $(dP)(dr)(2\pi r)$. Therefore, we have

$$dF = 2\pi(dx)d(r\tau) = (dP)(dr)(2\pi r) \tag{1.3}$$

Dividing Eq. 1.3 by $dxdr$ and using Eq. 1.1, we obtain

$$r\frac{dP}{dx} = \frac{d(r\tau)}{dr} = \frac{-d(r\mu\frac{du}{dr})}{dr} \tag{1.4}$$

Note that the water pressure here is assumed to be uniform along the r direction. This is a common and reasonable assumption used for studying fluid flow in a capillary tube with a radius much smaller than its length. The above equation can be solved for velocity gradient, $\frac{du}{dr}$, for a constant pressure gradient, $\frac{dP}{dx}$. Thus we have

$$\frac{r^2}{2}\frac{dP}{dx} + C = -r\mu\frac{du}{dr} \tag{1.5}$$

where C is a constant that is equal to zero as a result of the following boundary condition (or the symmetry condition):

$$\left.\frac{du}{dr}\right|_{r=0} = 0 \tag{1.6}$$

Then Eq. 1.5 can be further simplified as

$$-\frac{du}{dr} = \left(\frac{dP}{dx}\right)\left(\frac{r}{2\mu}\right) \tag{1.7}$$

Based on the non-slip conditions on the solid surface of the capillary tube ($u = 0$ at $r = R$), we can obtain the solution to Eq. 1.7 as

$$u(r) = \frac{dP}{dx}\int_{R}^{r}\frac{r}{2\mu}dr = \frac{1}{4\mu}\frac{dP}{dx}\left[R^2 - r^2\right] \tag{1.8}$$

The above equation gives the velocity distribution along the radial direction within the capillary tube. The average water flux across the cross section of the tube is then determined by

$$q_x = \frac{\int_0^R u(r)(dr)(2\pi r)}{\pi R^2} = -\frac{1}{\mu}\frac{R^2}{8}\frac{dP}{dx} = -\frac{\rho g}{\mu}\frac{R^2}{8}\frac{d(P/\rho g)}{dx} \tag{1.9}$$

where ρ is water density and considered to be a constant and g is gravitational acceleration. The above equation is in fact equivalent to the well-known Poiseuille equation. It was derived for a horizontal capillary tube. For an arbitrarily orientated straight capillary tube, the similar derivation procedure will give rise to the following relation:

$$q_x = -\frac{\rho g}{\mu}\frac{R^2}{8}\frac{dH}{dx} \tag{1.10}$$

where H is called hydraulic head and given by

$$H = z + \frac{P}{\rho g} \tag{1.11}$$

where z is elevation of the cross-section's center of capillary tube and positive in the upward direction.

If one, like Henry Darcy, had foreseen the (conceptual) similarity between water flow in porous media and capillary tubes, he, based on the above results, could have immediately recognized two important points. The first is, of course, Darcy's Law that water flux or flow rate is proportional to hydraulic gradient. The second is that the proportionality, called hydraulic conductivity, consists of two parts related to fluid properties and medium properties, respectively. To the best of our knowledge, the second point seems to be first made by M. King Hubbert (1903–1989) in his master piece regarding groundwater flow (Hubbert 1940). Thus, Eq. 1.10 can be written as

$$q_x = -\frac{\rho g}{\mu} k \frac{dH}{dx} \tag{1.12}$$

where k is called permeability that is determined by geometry of pore space or flow paths, like the radius of capillary tube (R) in Eq. 1.10, and thus an intrinsic property of a porous medium.

Any conceptualization or model is an approximation of the reality. It must be confirmed by experimental results. Darcy's sand column experiments were the necessary step for discovering Darcy's law. Nevertheless, subsurface problems are generally very complex in terms of medium geometry and the involved physical processes. In this case, it would not be possible, or practically necessary, to consider all the details related to the problems of interest. The use of a simplified concept to capture the essence of physics in a real system and leaving the unconsidered complexity to parameters that need to be determined by test data seem to be an effective way to develop practically useful theories or models. In Eq. 1.12, the unconsidered complexity by using a capillary tube to conceptualize pore space of a porous medium is included in permeability k that generally needs to be experimentally determined. Readers will find the similar approaches used for developing theories or models in other chapters as well in this book.

1.3 Generalized Darcy's Law for Water Flow in Low-Permeability Media

In his comments on Darcy's law, Freeze (1994) indicated that "it is fortuitous that God made life so simple, for if these relations (linear laws including Darcy's law) were nonlinear (if they had squared terms or cross products, for example), most of methods used by scientists to analyze the flow of water, heat, and electricity would be much messier." In the past two decades, significant progress has been made in understanding and applying nonlinearity sciences, including fractal geometry and chaotic systems. In fact, more and more scientists would agree on that God must be a nonlinearity expert who created so many beautiful things around us with nonlinearity, such as trees and flow-path patterns in river basins; most natural phenomena are actually associated with nonlinear processes, rather than linear processes. All the generalized physical laws, presented in this book, contain some nonlinear terms.

While significant attention has been historically given to flow processes in relatively high-permeability formations such as aquifers and conventional oil and gas reservoirs, fluid flow in low-permeability media (such as shale) becomes more and more important for a number of contemporary practical applications. For example, several countries have considered shale formations as potential host rocks for geological disposal of high-level radioactive nuclear wastes because shale has low permeability (on the order of 10–100 nano Darcy), low diffusion coefficient, high retention capacity for radionuclides, and capability to self-seal fractures (Tsang et al. 2012). Shale formations are also cap rocks for geological formations where supercritical CO_2 is stored for the purpose of CO_2 sequestration. Slow brine flow through the cap rock could be an important process that needs to be considered for managing pressure buildup owing to injection of CO_2 into storage formations below the cap rocks (Zhou et al. 2008). Nowadays, unconventional energy resources, including shale oil and shale gas, become an important part of recoverable hydrocarbon energy resources in the oil and gas industry. The recovery of these resources requires improved understanding of and modeling approaches for fluid flow within shale formations under different conditions.

The porous media used in Darcy's tests that led to the discovery of Darcy's law are sands, not shale or other low-permeability media. It has been well documented that Darcy's law is not adequate for the latter media. To provide some background for generalizing Darcy's law for the low-permeability media, this section first reviews the currently available relationships between water flux and hydraulic gradient and then develops an improved relationship that can capture water flow behavior in both high- and low-permeability media.

For one-dimensional flow systems, Darcy's law can be rewritten as

$$q_x = -Ki\mathrm{sgn}(\mathbf{i}_x) \tag{1.13}$$

where q_x is water flux with the magnitude of q, K is hydraulic conductivity and $\boldsymbol{i_x}$ is hydraulic gradient with a magnitude of i. The sgn($\boldsymbol{i_x}$) is the sign function of the hydraulic gradient and equals to 1, −1 and 0 for the positive, negative and zero gradients, respectively. The hydraulic conductivity is related to permeability, k, by

$$K = \frac{k\rho g}{\mu} \tag{1.14}$$

Non-Darcian flow is characterized by nonlinear relationship between water flux and hydraulic gradient (Liu et al. 2015). It is a result of strong solid–water interaction (e.g., Miller and Low 1963). The term "solid–liquid interaction" refers to combined effects of a variety of forces between molecules in solid and aqueous phases, including van der Waals forces. This interaction likely occurs in a thin fluid boundary layer close to solid surface where fluid properties are very different from those without the interaction. The thickness of the boundary layer is negligible in high-permeability media with relatively large pore sizes, but may be comparable to nanometer pore sizes in low-permeability media, such as shale formations. Thus, Darcy's law, based on the viscous laminar flow mechanism only, holds for high-permeability media, and is not adequate for fully describing liquid flow in low-permeability media.

The impact of solid–water interaction on water flow behavior has been demonstrated by a number of studies on nanoscale fluid flow based on molecular dynamics (MD) simulations (e.g., Chen et al. 2008; Ma et al. 2010; Farrow et al. 2011) and by experimental observations for water flow in microtubes (Xu et al.

2007). These studies generally show that water properties and flow processes at that scale could be significantly different from those in less fine-grained materials. For example, the density of water near the solid surface of grain material is generally much higher than its bulk value. Also, the resistance to water flow at the nanoscale is dominated by friction between fluid and solid surfaces, and less so by internal friction between fluid layers, because of the small pore sizes (Chen et al. 2008). Several MD simulation results show that the flow rate of water through a nano-tube is a nonlinear function of shearing stress (equivalent to hydraulic gradient for steady-state flow) (e.g., Chen et al. 2008; Ma et al. 2010; Farrow et al. 2011), which is consistent with non-Darcian behavior observed from laboratory measurements (e.g., Miller and Low 1963). Most recently, the MD simulation results of He et al. (2015) offer further insights, although their study is about gas transport in nanometer pores, rather than liquid. They simulated gas transport between two parallel carbon plates and within a carbon nanotube. The distance between the two plates (4.4 nm) is the same as nanotube diameter. They found that average gas velocity for the carbon-plate case is linearly related to the forces imposed to the gas molecules along the flow direction, which is consistent with Darcy's law at the macroscopic scale. That relationship becomes nonlinear for the nanotube case, an indication of non-Darcian flow behavior. This is because the latter is subject to a stronger field of intermolecular forces; the nanotube case corresponds to a larger ratio of solid surface area to the associated fluid volume.

Xu et al. (2007) also experimentally investigated the relationship between flux of deionized water and hydraulic gradient in individual microtubes with diameters ranging from 2 to 30 μm. They demonstrated that water flow in microtubes with diameters larger than 16 μm is consistent with Darcy's law, but not for smaller diameters. In the latter cases, the solid–water interaction is relatively strong and consequently the relationship between water flux and hydraulic gradient becomes nonlinear.

A number of researchers have proposed parametric formulations for describing the non-linear relationship between water flux and hydraulic gradient. Hansbo (1960, 2001) reported that water flux in a low-permeability clay soil could not be described by Darcy's law, but is proportional to a power function of the hydraulic gradient when the gradient is less than a critical value, whereupon the relationship between water flux and gradient becomes linear for larger gradient values. Consequently, Hansbo's (1960, 2001) proposed the following relationship:

$$q_x = -K' i^N \operatorname{sgn}(\mathbf{i}_x) \quad \text{for } i \leq i_1 \tag{1.15-1}$$

$$q_x = -K' N i_1^{N-1} (i - I) \operatorname{sgn}(\mathbf{i}_x) \quad \text{for } i \geq i_1 \tag{1.15-2}$$

$$i_1 = \frac{IN}{(N-1)} \tag{1.15-3}$$

The formulation of Hansbo (1960, 2001) includes three parameters K' (m/s), N (-) and I (m/m). Note that K' herein is not the hydraulic conductivity because

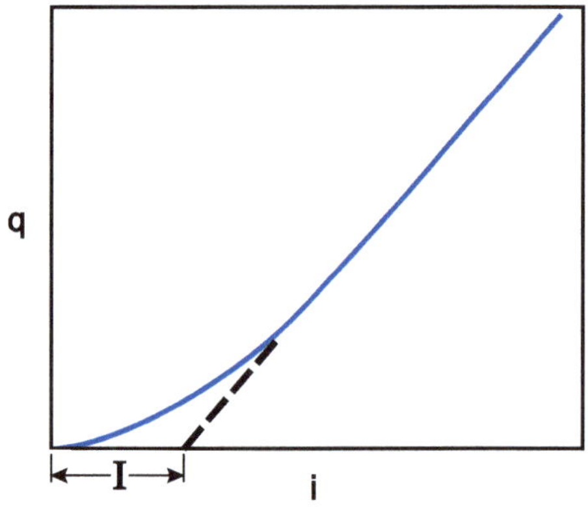

Fig. 1.5 Definition of threshold hydraulic gradient

Eq. 1.15-1 is not a linear function between water flux and hydraulic gradient i. Parameter I is called threshold gradient in this book and refers to the intersection between the i axis and the linear part of the relationship (Fig. 1.5). Hansbo (1960, 2001) explained the observed water-flow behavior by positing that a certain hydraulic gradient is required to overcome the maximum binding energy of mobile pore water. From their experiment results, Miller and Low (1963) also found the existence of a hydraulic gradient below which water is essentially immobile.

Hansbo (1960, 2001) demonstrated that Eq. 1.15 can fit related experimental observations and developed, based on Eq. 1.15, a theoretical approach to dealing with clay consolidation processes. As indicated by Swartzendruber (1961), however, Eq. 1.15 consists of two separate mathematical expressions and three related parameters and therefore is difficult to use in practice. To overcome this, Swartzendruber (1961), after analyzing several data sets for water flow in clay media, proposed a new version of the modified Darcy's law based on a relation for dq/di:

$$\frac{dq}{di} = K(1 - e^{-i/I}) \tag{1.16}$$

For a large value of hydraulic gradient i, dq/di approaches a constant K that is hydraulic conductivity. Integrating Eq. 1.16 and using the condition that $q = 0$ at $i = 0$, Swartzendruber (1961) obtains

$$q = K[i - I(1 - e^{-i/I})] \tag{1.17-1}$$

or

$$q_x = -K[i - I(1 - e^{-i/I})]\mathrm{sgn}(\mathbf{i}_x) \tag{1.17-2}$$

There are two parameters K and I in Eq. 1.17. Compared with the common form of Darcy's law, it contains only one additional parameter (I). The equation of

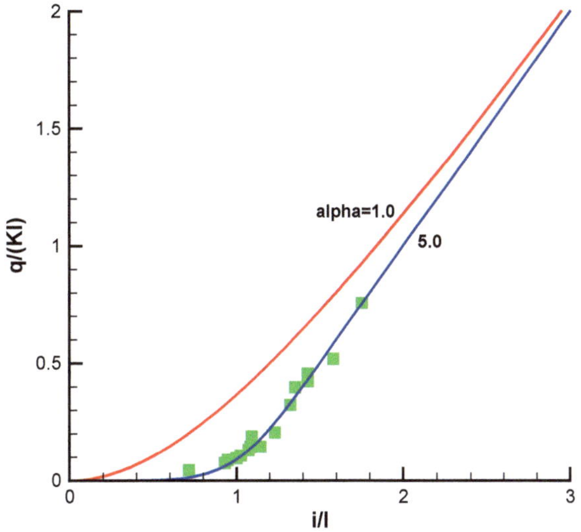

Fig. 1.6 Comparisons between Eq. 1.21 with two different α values and data of Cui et al. (2008) for the constant-volume tests (Liu and Birkholzer 2013) (reproduced by permission of Elsevier). Note that Swartzendruber equation (Eq. 1.17) corresponds to the curve for $\alpha = 1$

Swartzendruber (1961) had been evaluated with a number of data sets and satisfactory agreements were generally obtained (Swartzendruber 1961; Blecker 1970). However, Fig. 1.6 clearly demonstrates the deviation between results calculated from Eq. 1.17 and a data set collected by Cui et al. (2008) under the unsaturated condition. Thus, Eq. 1.17 cannot capture the full range of non-Darcian flow behavior under different conditions. (Note that under unsaturated conditions, K in Eq. 1.17 and other related equations corresponds to the unsaturated conductivity that is defined as the hydraulic conductivity multiplied by the relative permeability for water flow.)

Another commonly used flux-gradient relationship is given as (e.g., Bear 1979):

$$q_x = 0 \quad \text{for} \quad i \leq I \tag{1.18-1}$$

$$q_x = -K(i - I)\text{sgn}(\mathbf{i}_x) \quad \text{for} \quad i \geq I \tag{1.18-2}$$

The above equation also involves only two parameters (K and I) and is mathematically simpler than other relationships. However, it cannot adequately capture the non-Darcian flow behavior (or nonlinear flux-gradient relationship) (e.g., Swartzendruber 1961; Blecker 1970), because flux is not always zero at low i values in reality. Therefore, Eq. 1.18 should be applied only when i is large such that the flux-gradient relationship in the low i range is not important anymore. It is also of interest to note that Eq. 1.18-2 is a limiting case of Eq. 1.17 for $i/I \rightarrow \infty$.

Zou (1996) proposed a nonlinear flux-gradient relationship based on an assumption that the activation energy of pore water is not only variable with the distance from the solid particle surface, but also with the flow velocity of pore water. His model includes several empirical parameters and is able to fit several data sets that show nonlinear flux-gradient relationships at low hydraulic gradients. His flux-gradient relationship is:

$$q = \frac{Ki}{1 + b\exp(-ci)} \tag{1.19}$$

where b and c are empirical parameters. However, this equation, for $i \rightarrow \infty$, gives a linear flux-gradient relation going through the origin in Fig. 1.5, or is reduced to the exact form of Darcy's law. In other words, Eq. 1.19 cannot capture the effect of threshold gradient on the flux-gradient relation for large gradient i, while this effect has been often observed from test results, as shown in Fig. 1.6.

From the above discussions, it is clear that a more general relationship between water flux and hydraulic gradient is required. Most recently, Liu and Birkholzer (2013) proposed to generalize the Swartzendruber's (1961) relationship using:

$$\frac{dq}{di} = K\left(1 - e^{-\left(\frac{i}{I^*}\right)^{\alpha}}\right) \tag{1.20}$$

where α is a non-negative constant, and I^* is a parameter related to α and I and will be defined in Eq. 1.22-1. For $\alpha = 1$, Eq. 1.20 is reduced to Eq. 1.16. For $\alpha \rightarrow \infty$, we have $\frac{dq}{di} \rightarrow 0$ for $\frac{i}{I^*} < 1$ and $\frac{dq}{di} \rightarrow K$ for $\frac{i}{I^*} > 1$. In this case, Eq. 1.20 essentially represents the flux-gradient behavior given in Eq. 1.18, as will be demonstrated in Fig. 1.7. Thus, with one more parameter (α), Eq. 1.20 can capture a relatively large range of non-Darcian flow behavior.

Integrating Eq. 1.20 and again using the condition of $q = 0$ at $i = 0$, Liu and Birkholzer (2013) obtains

$$q = K\left[i - \frac{I}{\gamma\left(\frac{1}{\alpha}\right)}\gamma\left(\frac{1}{\alpha}, \left(\frac{i}{I^*}\right)^{\alpha}\right)\right] \tag{1.21-1}$$

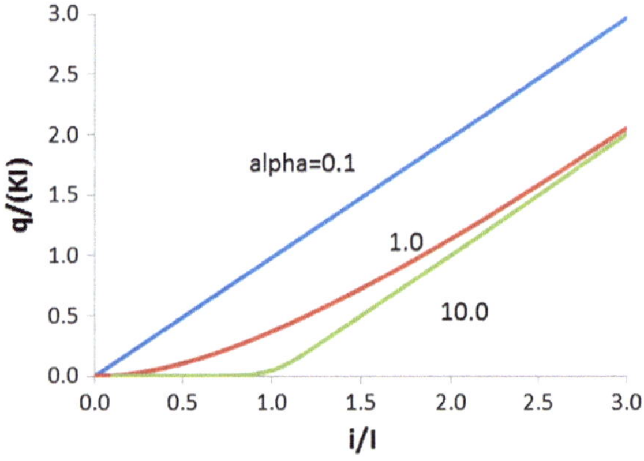

Fig. 1.7 Relationship between normalized flux and gradient calculated from Eq. 1.21-1 for $\alpha = 0.1$, 1.0 and 10.0

or

$$q_x = -K\left[i - \frac{I}{\gamma\left(\frac{1}{\alpha}\right)}\gamma\left(\frac{1}{\alpha}, \left(\frac{i}{I^*}\right)^{\alpha}\right)\right]\mathrm{sgn}(\mathbf{i}_x) \tag{1.21-2}$$

where

$$I = \frac{I^*}{\alpha}\gamma\left(\frac{1}{\alpha}\right) \tag{1.22-1}$$

and γ refers to Gamma functions

$$\gamma(a,x) = \int_0^x t^{a-1}\mathrm{e}^{-t}dt \tag{1.22-2}$$

$$\gamma(a) = \int_0^\infty t^{a-1}\mathrm{e}^{-t}dt \tag{1.22-3}$$

In the above equation, t is a dummy variable. (Note that t also refers to time in Sect. 1.7.3.)

Figure 1.6 shows a good agreement between Eq. 1.21 with $\alpha = 5$ and the data from Cui et al. (2008). Note that Cui et al. (2008) reported flux-gradient data for three capillary pressure values. Each capillary pressure has its own I value. Different I values are used in Fig. 1.6 for different capillary pressure when calculating i/I and dimensionless flux defined as $q/(KI)$. The determination of parameters I and K is discussed in the next section.

Equation 1.21 is the generalized Darcy's law (Liu and Birkholzer 2013), although it should be better considered a phenomenological model at this point. (Note that Darcy's law was discovered as a phenomenological law as well.) For the threshold hydraulic gradient $I = 0$, Eq. 1.21 is reduced to Darcy's law. Thus, Darcy's law and two important modified versions of Darcy's law (Eqs. 1.17 and 1.18) are special cases of Eq. 1.21. The generalized Darcy's law includes three parameters, hydraulic conductivity K, threshold hydraulic gradient I, and parameter α. They are mainly characterized by the average pore size, degree of solid–liquid interaction, and pore-size distribution, respectively. These parameters and their correlations will be further discussed in the next two sections. Figure 1.7 shows relations between the normalized flux and the gradient calculated from Eq. 1.21-1 for $\alpha = 0.1$, 1.0 and 10.0, respectively, demonstrating that Eq. 1.21 covers the full range of the currently known flux-gradient relationships.

1.4 Correlation Between Permeability and the Threshold Gradient

A correlation between permeability (or average pore size) and the threshold hydraulic gradient *I* for low-permeability materials is expected, because non-Darcian flow behavior is closely related to the pore size. A smaller pore size corresponds to a larger ratio of solid-liquid interfacial area to liquid volume, or a stronger solid–liquid interaction. We will demonstrate such a relationship with a number of data sets of permeability as a function of threshold hydraulic gradient (Fig. 1.8) in this section.

Miller and Low (1963) presented results of laboratory experiments for Wyoming bentonite soil samples (with a length of 2.5 cm and diameter of 1 cm). Under the steady-state flow condition, they measured water flow rates through soil samples for different hydraulic gradients across the samples. Then water flux is calculated as the corresponding water flow rate divided by the sample cross-sectional area. The permeability values are estimated here using the slope of the linear part of water flux versus gradient curves. When the number of data points corresponding to the linear part is limited, we use a straight line through the two data points with the largest gradients to approximate the linear part. Note that our definition of the threshold gradient is different from that by Miller and Low (1963) whose threshold gradient is the gradient below which water flow could not be detected. Our threshold gradient *I* is used to describe the intersection between the gradient axis and the linear part of the flux-gradient relationship (Fig. 1.5). As an example,

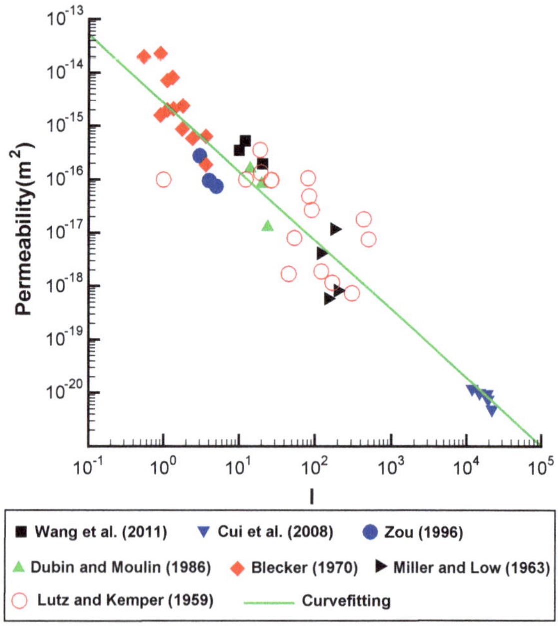

Fig. 1.8 Correlation between permeability and threshold hydraulic gradient (Liu and Birkholzer 2013) (reproduced by permission of Elsevier)

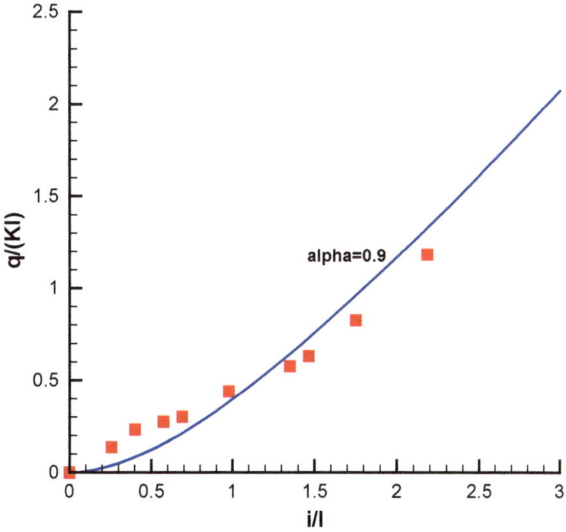

Fig. 1.9 Match of Eq. 1.21 with test data for a Na-clay soil from Miller and Low (1963) (Liu and Birkholzer 2013) (reproduced by permission of Elsevier)

Fig. 1.9 shows a match of Eq. 1.21 with data from one of the experiments conducted by Miller and Low (1963).

Blecker (1970) documented laboratory test results for relationships between permeability and hydraulic gradient for clay soils from Brolliar and Springerville soil series of northern Arizona, USA. His experimental setup is similar to the one used by Miller and Low (1963). The soil samples were compacted to have different soil densities for a given soil type. A larger density generally corresponds to a smaller porosity and average pore size and therefore to a smaller permeability. Blecker (1970) showed a good agreement between his data and Eq. 1.17 and also reported values for fitted permeability and I that are presented in Fig. 1.8.

Consolidation of clays is important for some civil-engineering applications and closely related to drainage process that is largely determined by clay permeability under different stress and hydraulic conditions (Hansbo 2001). For this purpose, Dubin and Moulin (1986) measured the flux-gradient curves for Saint Herblain clay. From these curves, values for permeability and the threshold gradient are obtained and presented in Fig. 1.8. To evaluate his relationship for water flux and hydraulic gradient (Eq. 1.19), Zou (1996) also presented a data set for several soil types. Figure 1.8 only shows his data points for clay soil because no significant non-Darcian behavior is observed from other soils in his paper.

Wang et al. (2011) reported core test data for rock samples collected from low-permeability Daqing oilfield, China, including steady-state water flow rate as a function of hydraulic gradient imposed across the samples under the saturated condition. The non-Darcian flow behavior is represented by the ratio of $(\frac{q}{i})/K$; this ratio should be equal to one for Darcian flow (Eq. 1.13). Based on Eq. 1.17, the ratio $(\frac{q}{i})/K$ can be expressed as

$$\frac{q/i}{K} = 1 - \frac{1 - e^{-\frac{i}{I}}}{i/I} \tag{1.23}$$

The above equation is used to fit the data sets of Wang et al. (2011), as shown in Fig. 1.10. The fitting is achieved by adjusting parameter I for a given rock sample. The fitted I values are 20.4, 12.0 and 10.0 for rock samples with permeability values of 0.195, 0.347 and 0.524 mD, respectively (Liu and Birkholzer 2013). Note that 1 mD is equal to 9.87E−16 m^2. The match between Eq. 1.23 and the data is reasonable, indicating that Eq. 1.17 is valid for describing flux-gradient relationships for these rock samples. Clearly, Fig. 1.10 shows that $(\frac{q}{i})/K$ is not constant and less than one for small i/I values. In other words, it is more difficult for fluid to flow for small i/I values, an indication of non-Darcian flow in low-permeability media.

Lutz and Kemper (1959) reported the impact of clay–water interaction on hydraulic properties of clay soils (including Utah bentonite, halloysite, Bladen clay, and Wyoming bentonite). The tests were conducted under the saturated condition using fluids with different electrolyte concentrations; soil sample volumes were fixed by sample holders. For a given soil and hydraulic gradient, a low concentration generally corresponds to a low water flow rate or low permeability, because swelling reduces average pore size for a given soil sample volume. The steady-state water flow rate was monitored as a function of the gradient in their tests. The threshold gradient values for all these tests were calculated by Swartzendruber (1961) and used in Fig. 1.8. The data of Lutz and Kemper (1959) is consistent with the general trend in the figure, indicating that the impact of soil swelling, to a large extent, has been included in permeability values because permeability is a strong function of swelling (or shrinkage).

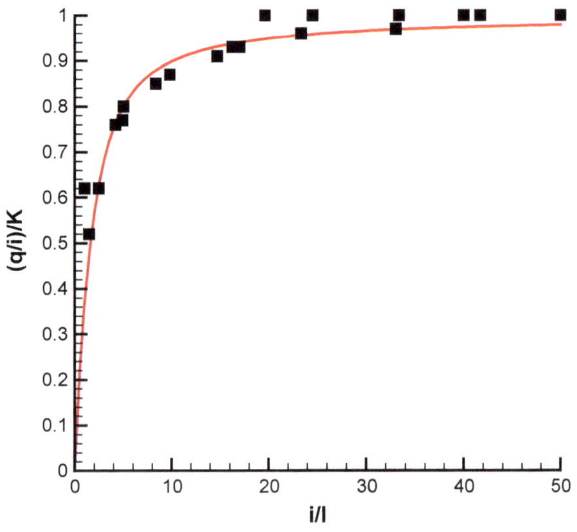

Fig. 1.10 Match of Eq. 1.23 with test data of Wang et al. (2011) (Liu and Birkholzer 2013) (reproduced by permission of Elsevier)

All the data sets discussed above have been collected under the saturated condition. Cui et al. (2008) reported measurements of unsaturated hydraulic conductivity for a sand-bentonite mixture compacted in a metallic cylinder (50 mm in inner diameter, 250 mm high). The instantaneous profile method was employed to determine the unsaturated hydraulic conductivity (Daniel 1982). The bottom of the vertical test cell was connected to a water reservoir, and the upper end to the atmosphere. Under transient and upward water-flow conditions, vertical distributions of capillary pressure were directly measured as a function of time at several locations along the column. Then the corresponding vertical distributions of water content were estimated from the relationship between water content and capillary pressure that was independently measured under constant volume conditions. Based on these time-dependent vertical distributions and water mass balance at each location within the soil column, they were able to estimate water flux at those locations as a function of capillary pressure and hydraulic gradient. Details of the instantaneous profile method can be found in Daniel (1982) and Cui et al. (2008). The linear part of the flux-gradient relationship for a given capillary pressure is approximated by a straight line through the two points with the largest hydraulic gradients. Consequently, the threshold gradient I and the unsaturated permeability (that corresponds to the slope of the straight line mentioned above) can be obtained for each capillary pressure. Figure 1.6 shows the match between Eq. 1.21 and the data set.

As shown in Fig. 1.8, the data sets discussed above can be reasonably fitted by an empirical relationship between I and permeability k (m^2) (Liu and Birkholzer 2013)

$$I = A'k^{B'} \tag{1.24}$$

with $A' = 4.0 \times 10^{12}$ and $B' = -0.78$. This may imply the existence of a universal relationship between I and permeability, given the fact that these data sets were collected by different researchers for different types of low-permeability media.

When permeability for a given medium is known, Eq. 1.24 (or Fig. 1.8) can be used to evaluate the relative importance of non-Darcian liquid behavior (Liu and Birkholzer 2013). As previously indicated, a larger I gives rise to a stronger non-Darcian flow behavior. For example, non-Darcian behavior may be negligible for modeling liquid flow in porous media with permeability values larger than 10^{-13} m^2 because the corresponding I values are small, as shown in Fig. 1.8. When data are not available for estimating site-specific values for A' and B', Eq. 1.24 can be employed for estimating parameter I based on more easily obtained permeability values. Equation 1.24 is also useful for providing a criterion for permeability to be accurately measured. For example, if the hydraulic gradient head used in tests is less than the value calculated from Eq. 1.24, the measured permeability will depend on the gradient and does not correspond to the correct value, which will be further discussed later in this chapter. The data for unsaturated flow seems to follow the same relationship as that for saturated flow when permeability is replaced by unsaturated permeability (Fig. 1.8). The unsaturated permeability is related to sizes of pores occupied by water under unsaturated conditions.

Parameters in Eq. 1.21 depend on the electrolyte concentration of liquid solution that flows through porous media and on its temperature. For example, Swartzendruber (1961) analyzed the data of Lutz and Kemper (1959) that were collected for solutions (with different electrolyte concentrations) flowing through clay soils and demonstrated that the threshold gradient decreases with the concentration. As previously indicated, the impact of the concentration is partially considered herein through the correlation between permeability and I (Eq. 1.24). For example, with increasing solution concentration, permeability will increase because of shrinkage and thus parameter I will decrease based on Eq. 1.24. This is consistent with the analysis results of Swartzendruber (1961). On the other hand, osmotic flow can be important for low-permeability materials under certain conditions, depending on solute concentration gradient and osmotic efficiency (Barbour et al. 1991; Kang 2008). For the experiments discussed in this section, osmotic flow, however, is not significant under those test conditions, because no significant water flow was observed for zero-pressure gradient (under which water flow is controlled by osmosis). Therefore, we follow the authors who initially collected and/or analyzed the data sets discussed in this section and ignore osmotic flow when analyzing these data sets.

There are very limited studies of the temperature impact on non-Darcian behavior in the literature. Miller and Low (1963) experimentally found that increasing temperature reduces their threshold hydraulic gradient (below which water flow could not be detected). They speculated that an increased temperature may weaken the forces between water molecules and clay surface (or solid–liquid interaction). More experimental and theoretical studies are needed to fully understand temperature dependence of related parameters in Eq. 1.21.

1.5 Relationship Between Parameter α and Pore Size Distribution

To make sure that the generalized Darcy's law (Eq. 1.21) is physically valid, we require Eq. 1.20 to mathematically satisfy certain conditions (or constraints). The first condition is that the magnitude of water flux q always does not decrease with increasing gradient i, or

$$\frac{dq}{di} = K(1 - \mathrm{e}^{-\left(\frac{i}{I^*}\right)^{\alpha}}) \geq 0 \tag{1.25}$$

That is equivalent to

$$\left(\frac{i}{I^*}\right)^{\alpha} \geq 0 \tag{1.26}$$

Obviously, the above equation is always satisfied. That is one of the motivations to derive the generalized Darcy's law based on Eq. 1.20.

The experimental observations discussed in Sect. 1.4 indicate that dq/di always increases with i and reaches its maximum value when the relationship between q and i becomes linear. Thus, Eq. 1.20 must satisfy this condition, or

$$\frac{d^2q}{di^2} \geq 0 \tag{1.27}$$

Based on Eq. 1.20, the above equation becomes

$$\alpha \frac{K}{I^*} \left(\frac{i}{I^*} \right)^{\alpha - 1} e^{-\left(\frac{i}{I^*}\right)^{\alpha}} \geq 0 \tag{1.28}$$

That leads to

$$\alpha \geq 0 \tag{1.29}$$

because all the other terms in Eq. 1.28 are not negative. Thus, parameter α only takes non-negative values, as indicated in Sect. 1.3.

Parameter α largely characterizes how smoothly transition occurs from the nonlinear region to the linear region in a $q(i)$ curve (Fig. 1.7). For example, for $\alpha \to \infty$, the generalized Darcy's law is reduced to Eq. 1.18 in which dq/di becomes discontinuous at $i = I$. On the other hand, when $\alpha \to 0$, dq/di becomes a constant of $K(1 - e^{-1})$ and eventually approaches K for $i \to \infty$. In other words, the transition of the $q(i)$ curve is extremely smooth. Thus, the smaller the α value, the smoother the transition from the nonlinear to the linear region in a $q(i)$ curve.

Parameter α is related to the pore size distribution because pores with different sizes correspond to different degrees of non-Darcian flow behavior, as a result of different degrees of solid–liquid interaction. It should be easy to understand that a broader pore-size distribution corresponds to a more gradual (or smoother) transition from the nonlinear region to the linear region in a $q(i)$ curve, and therefore to a smaller α value.

To demonstrate the impact of pore size distribution on the shape of a $q(i)$ curve, let's consider two ideal porous media that have the same hydraulic conductivity K and threshold hydraulic gradient I, but different pore size distributions. Just for the demonstration purpose, we assume that the non-Darcian flow behavior is described by Eq. 1.18 in which q is zero for i less than or equal to I and then linearly increases with i. Furthermore, the first porous medium has a uniform pore size and corresponds to curve L_1 in Fig. 1.11. The second porous medium is a layered system with flow direction along the interface between its two layers with uniform thicknesses. One layer has a larger (and uniform) pore size than the first porous medium, such that Darcian flow occurs within the layer (L_2 in Fig. 1.11). The layer contributes to half of the total flux for the medium; the total flux is defined as the total flow rate divided by the cross-sectional area of the layered

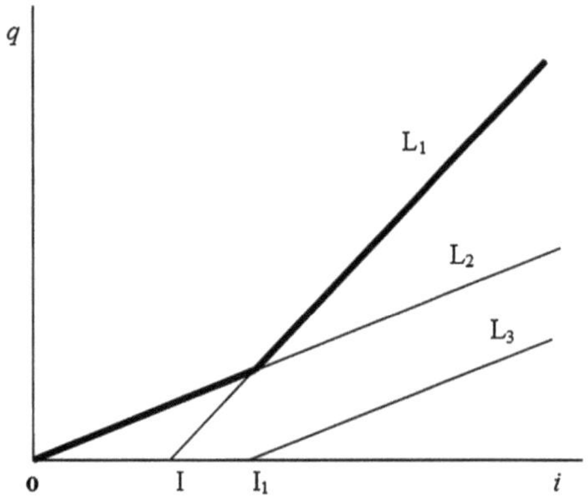

Fig. 1.11 Demonstration of impact of pore size distribution on the shape of the $q(i)$ curve

system. The second layer has a smaller (and uniform) pore size than the first porous medium and thus the threshold hydraulic gradient I_1 is larger than I (L_3 in Fig. 1.11). This layer contributes to the second half of the total flux for the layered system. (Obviously, the second layer should be much thicker than the first layer because they have the same contribution to the total liquid flux for a given hydraulic gradient larger than I_1.) For the second porous medium, the relationship between the magnitude of total flux q and hydraulic gradient i is the additive combination of L_2 and L_3 in Fig. 1.11. The thick black line (with two segments from L_2 and L_1 respectively) is the $q(i)$ curve for the second medium (the layered system). Note that I_1 is the same as the hydraulic gradient at the intersection between L_1 and L_2, because the two ideal porous media have the same threshold gradient (I).

The second porous medium (the layered system) has a broader (yet discrete) pore size distribution than the first medium. As a result, the q changes more gradually with hydraulic gradient i than the first medium. For the purpose of demonstrating the concept, we have considered an extremely simple case in this section. The conclusion, however, holds in a more realistic case in which a continuous pore size distribution would generally result in a continuous slope (dq/di) for a $q(i)$ curve.

1.6 Multidimensional and Anisotropic Cases

The generalized Darcy's law was developed for one-dimensional water flow, as discussed in Sect. 1.3. In many studies, we need to investigate multidimensional water-flow processes. For that purpose, Liu (2014) extends Eq. 1.21 to three-dimensional, homogeneous and isotropic media:

$$\mathbf{q} = -K\left[i - \frac{I}{\gamma\left(\frac{1}{\alpha}\right)}\gamma\left(\frac{1}{\alpha}, \left(\frac{i}{I^*}\right)^{\alpha}\right)\right]\mathbf{n_i} \tag{1.30}$$

where **q** (m/s) and $\mathbf{n_i}$ (-) are water flux vector and unit vector for hydraulic gradient, respectively. The relationship between I and I^* is given by Eq. 1.21-1. When the threshold hydraulic gradient I approaches zero, Eq. 1.30 is reduced to the commonly used form of Darcy's law. It should be emphasized that Eq. 1.30 is an empirical relationship and developed based on one-dimensional test results. It is logical to directly extend Eq. 1.21 to Eq. 1.30 for homogeneous and isotropic cases because both equations are consistent with the one-dimensional test condition that flux and negative hydraulic gradient are along the same direction.

Shale formations are generally anisotropic, as a result of the existence of bedding structures in these formations. To model water flow in them, we need a formulation of water flux in anisotropic media. Because of the nonlinear feature of the non-Darcian flow, it is not a trivial task to theoretically relate flux-gradient relationships between different dimensions and/or between isotropic and anisotropic cases. Thus, Liu (2014) presents an approximate formulation of water flux in an anisotropic medium. For simplicity, Liu (2014) choose the spatial coordinate axes to be in the principal directions of the anisotropic medium (Bear 1979) such that one axis is perpendicular to the bedding plane. Then, Liu (2014) proposes the water-flux formulation for anisotropic media by extending Eq. 1.30:

$$\boldsymbol{q} = -\boldsymbol{K}\left[i - \frac{I}{\gamma\left(\frac{1}{\alpha}\right)}\gamma\left(\frac{1}{\alpha}, \left(\frac{i}{I^*}\right)^{\alpha}\right)\right]\boldsymbol{n}_i \tag{1.31}$$

$$I = \frac{\sqrt{(I_x i_x)^2 + (I_y i_y)^2 + (I_z i_z)^2}}{\sqrt{(i_x)^2 + (i_y)^2 + (i_z)^2}} \tag{1.32}$$

where $\boldsymbol{K}$ is the conductivity tensor, and subscripts x, y and z refer to components along three coordinate directions. Specifically, I_j ($j = x, y, z$) is the measured threshold hydraulic gradient in the jth direction from one-dimensional experiments. Because of the way the coordinate system is chosen, there are no cross terms in the conductivity tensor. It is also easy to show that the relation among the magnitude of hydraulic conductivity, conductivity components, and hydraulic gradient components has the same mathematical form as Eq. 1.32 if the relevant variables for the threshold hydraulic head in Eq. 1.32 are replaced by the corresponding variables for the conductivity.

The reasonableness of the approximation (Eqs. 1.31 and 1.32) can be demonstrated by its consistency with known results. For one-dimensional flow along the bedding direction (e.g., the x direction), Eq. 1.31 is reduced to Eq. 1.21 with $I = I_x$. Similarly, if the z direction is perpendicular to the bedding plane, we can also get the expected result with $I = I_z$ for one-dimensional flow along the z direction. For isotropic cases (or $I_x = I_y = I_z$), we can recover Eq. 1.30 from Eq. 1.31. Thus, Eq. 1.31 gives an approximate, yet practically reasonable way to model non-Darcian water flow in the bedding shale formations, while the exact formulation for water flux in anisotropic media is difficult to obtain (Liu 2014).

1.7 Case Studies

This section provides selected case studies to demonstrate the importance of non-Darcian flow behavior in several research areas associated with liquid flow in low-permeability media (Liu 2014; Liu et al. 2015).

1.7.1 Impact of Non-Darcian Flow on Performance of a Shale Repository for High-Level Nuclear Waste

Shale formations have been considered by several countries as potential host rock type for geological disposal of high-level radioactive waste (Tsang et al. 2012). One major concern for a shale repository is that hydromechanical perturbation caused by underground excavations during its construction stage could alter rock properties surrounding repository tunnels and thus could compromise the safety performance of the repository system. The perturbation generally results in an excavation damaged zone (EDZ) around the repository tunnels (where the nuclear waste packages will be stored) and access shaft used to access the repository from the ground surface, because of a redistribution of in situ stresses and rearrangement of rock structures (Tsang et al. 2012). Induced macro- and micro-fractures in the EDZ can increase the rock permeability by one or more orders of magnitude. Thus, the EDZ could act as a preferential flow path for advective transport. This is a very unfavorable condition because it can speed up radionuclide transport from failed waste packages toward the biosphere (Fig. 1.12). On the other hand, if water velocity in the EDZ was small, diffusion would be the dominant mechanism for radionuclide transport. In this case, the impact of the EDZ on radionuclide transport is minimal. Therefore, the relative importance of advection versus diffusion in the EDZ is a key issue for assessing the performance of a shale repository (Liu 2014).

Recently, Bianchi et al. (2013) conducted a comprehensive modeling study on the relative importance of advection versus diffusion within the EDZ. They developed a two-dimensional model of a generic repository including a single horizontal tunnel for waste emplacement, a vertical shaft, and a vertical cross

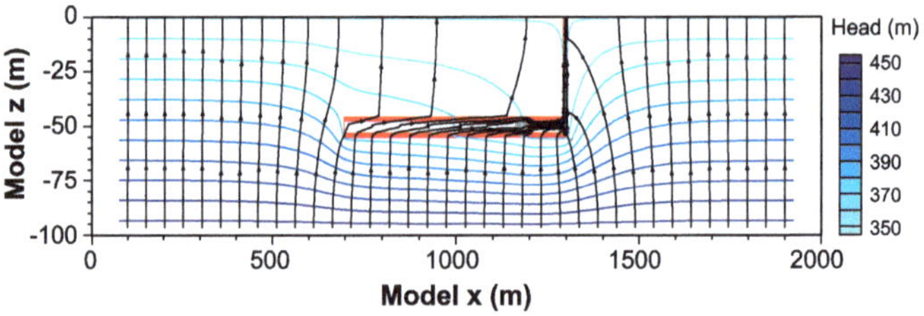

Fig. 1.12 Flow pattern for a two-dimensional clay repository (Liu 2014). The horizontal structure is a repository tunnel connected to the ground surface through a vertical access shaft

section of the host-rock formation. The longitudinal axis of the horizontal tunnel, which has a total length *Le* equal to 600 m, is located at $z = -50$ m (Fig. 1.12). ($z = 0$ on the ground surface.) The length of the model domain in the x direction is 2000 m, such that the left and right boundaries are sufficiently distant from the tunnel and the shaft. The total thickness of the host rock is 100 m, while the vertical extension of the shaft equals 50 m, from the end of the horizontal tunnel to the top boundary of the host formation. For their base case scenario, Bianchi et al. (2013) assumed a vertical hydraulic gradient of 1 m/m. This value is consistent with the hydraulic conditions of the Opalinus Clay at the Mont Terri site, Switzerland (Bianchi et al. 2013). The Mont Terri site has been intensively investigated as an analogue for a shale/clay repository. The values of the hydrogeological parameters assigned to the host rock and the repository components were also chosen to be realistic with respect to values from several sources (Table 1.1).

The base-case simulation results of Bianchi et al. (2013) are demonstrated in Fig. 1.12. The stream lines are generally in a vertical direction except within the EDZ surrounding the tunnel, and the water potential lines (each of which has the constant water head (potential)) are generally in the horizontal direction. The simulations did not consider the non-Darcian flow behavior. Bianchi et al. (2013) found that for a given ambient hydraulic gradient, the upward water flow below the tunnel is intercepted by the high-permeability EDZ and advection indeed becomes important within the EDZ. They also found that the advection within the EDZ is limited by the amount of water flowing into it from the surrounding host rock. This is expected because when there is no water flowing into the EDZ, water velocity within the EDZ would be zero, even though the EDZ permeability is high.

In his study, Liu (2014) further investigated the relative importance of advection versus diffusion within the EDZ by incorporating the non-Darcian flow behavior. The model setup and parameter values in Bianchi et al. (2013) are employed herein. To simplify the analysis procedure, Liu (2014) assumes that the EDZ permeability is infinite and the permeability of backfills in the tunnel is zero. Obviously, these assumptions generally lead to overestimation of the relative importance of advection within the EDZ. In this case, the EDZ has the same pressure head as the upper boundary in Fig. 1.12, because the EDZ is in equilibrium with the upper boundary as a result of EDZ's infinite permeability. The simulation results from Bianchi et al.

Table 1.1 Parameter values for the generic shale repository

Parameter	Value
Vertical thickness of the host rock formation	100 m
Total length of the repository in the x direction	600 m
Repository tunnel diameter	4.5 m
Thickness of the EDZ in the tunnel	2.4 m
Initial hydraulic gradient in the z direction	1 m/m
Host rock permeability	5E−20 m^2
Diffusion coefficient for EDZ fractures	1.08E−9 m^2/s
Fracture porosity within the EDZ	0.01

(2013) indicate that under steady state conditions, the hydraulic gradient in the zone below the tunnel is very close to the uniform distribution. The same distribution is assumed here. With these assumptions and approximations, the hydraulic behavior of the EDZ can be easily determined without using numerical modeling. The hydraulic gradient below the tunnel will be twice as large as the ambient condition (or $i = 2$ m/m). The EDZ intercepts all the upward flow from the region below the tunnel. The water flow rate (and velocity) will be linear along the x axis with zero value at the left end and the maximum value at the right end (Fig. 1.12).

The water flux into the EDZ from the lower portion of the clay formation in Fig. 1.12 is estimated with Eq. 1.18 (for simplicity) and for $i = 2$ m/m. Based on the mass balance, the maximum (horizontal) water flow rate within the EDZ is equal to the calculated vertical water flux, multiplied by the length of the tunnel and by the upper diameter of the EDZ. (The multiplication gives the area intercepting the upward water flow.) Then, the maximum pore velocity within the EDZ is determined as the flow rate divided by its cross-sectional area and by fracture porosity. Figure 1.13 shows the normalized pore velocity as a function of threshold gradient I. The normalized velocity there refers to the ratio of the maximum pore velocity within the EDZ to that for Darcian flow. The filled circle in Fig. 1.13 corresponds to the threshold hydraulic gradient $I = 2755$ m/m calculated from Eq. 1.24 using permeability $k = 5\text{E}{-20}\ \text{m}^2$ (Table 1.1). Obviously, the non-Darcian flow has a significant impact on the water velocity. For the problem under consideration with $I = 2755$ m/m, the water velocity corresponding to non-Darcian flow is more than three orders of magnitude smaller than that for the Darcian flow (Fig. 1.13).

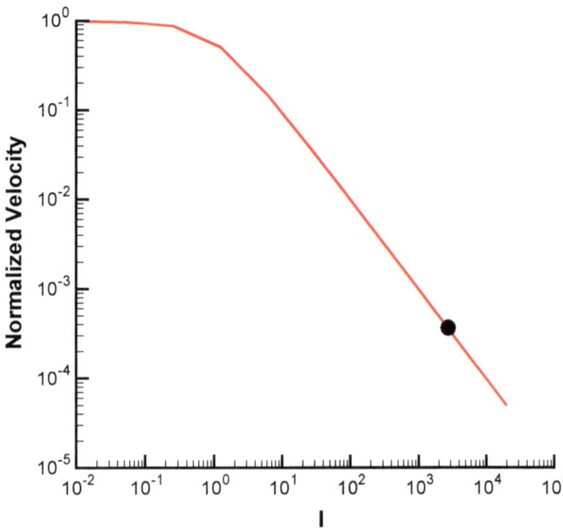

Fig. 1.13 The normalized pore velocity as a function of threshold gradient I (Liu 2014). The normalized velocity refers to the ratio of the maximum pore velocity within the EDZ to that in the Darcian-flow case

The relative importance of the advection within EDZ can be evaluated with the Peclet number, *Pe* (Bianchi et al. 2013; Liu 2014):

$$Pe = \frac{vLe}{D} \tag{1.33}$$

where v (m/s) is the maximum pore velocity, Le = 600 m is the length of the tunnel, and D (m^2/s) is the effective diffusion coefficient within fractures (Table 1.1). Note that for D here, unlike in porous media, we do not need to consider the tortuosity factor because transport paths in fractures are close to straight lines (Liu 2014). In general, *Pe* represents the ratio of the time for diffusion (Le^2/D) to the time for advection (Le/v). For porous media, $Pe < 1$ is considered to correspond to the solute-transport regime with dominant diffusion. This criterion, however, is too strict here because the matrix diffusion process can significantly enhance the role of the diffusion process in fractured rock (Liu et al. 2007), but is not considered in the expression for *Pe*. Nevertheless, the calculated *Pe* values for the problem given in Fig. 1.12 are 0.6 in the non-Darcian flow case and 1666.7 in the Darcian flow case, respectively. It is obvious that solute transport within the EDZ, as a result of non-Darcian flow behavior, is diffusion dominated, even though the EDZ permeability is assumed to be infinite. In other words, advection is not the dominant transport mechanism within the EDZ because the water flow rate into the EDZ is essentially eliminated by the non-Darcian flow behavior in a shale formation. Note that our conclusion holds only under conditions without intersections between the EDZ and the conductive geological structures that connect the high-permeability formations surrounding the host rock formation. This result is significant given the fact that EDZ's role in radionuclide transport has been a central issue in the performance assessment of a shale repository for nuclear waste.

1.7.2 *Influence of Non-Darcian Flow on Observed Relative Permeability*

Two-phase flow properties of low-permeability media, including relative permeability k_r, are necessary inputs to modeling flow processes in many practical applications. This subsection is going to show that if non-Darcian flow behavior is not considered, significant errors could occur in estimating liquid relative permeability from laboratory measurements (Liu 2014). The next chapter will be fully devoted to the two-phase flow process. If the reader is not familiar with the two-phase flow, he or she may skip this and next subsections and get back to them after going through the next chapter.

In general, the relative permeability can be related to water saturation by (van Genuchten 1980):

$$k_r = S_e^{1/2}\left\{\left[1-\left(1-S_e^{1/m}\right)^m\right]\right\}^2 \tag{1.34}$$

where m is a parameter related to pore size distribution and S_e is the effective saturation defined by

$$S_e = \frac{S - S_r}{1 - S_r} \tag{1.35}$$

where S and S_r are water saturation and residual saturation, respectively.

Since parameter m can be determined from the relatively easily measured water retention curve (or relation between capillary pressure and water saturation), Eq. 1.34 provides an efficient way to estimate relative permeability. The measurements of relative permeability are difficult and take long time. However, the validity of the van Genuchten (1980) relationship, as an empirical one, is an open question for the low-permeability media (such as shale rock) because there are very rare studies on the comparison between the relationship and measured relative-permeability data for the media. The major purpose of this subsection, as previously indicated, is to demonstrate that "measured" values for relative permeability are not true medium properties, but strongly depend on test conditions, as a result of non-Darcian flow behavior. By "true" relative permeability, we mean the relative permeability associated with the linear regime for the curve in Fig. 1.5.

To measure the relative permeability, the hydraulic gradient (i) is generally fixed, and then water flux (q) is measured under steady-state condition. In this case, the "measured" relative permeability k_{rM} can be calculated as

$$k_{rM} = \frac{q/i}{K} \tag{1.36}$$

For simplicity, Liu (2014) uses Eq. 1.18 to describe the non-Darcian flow behavior. Note that under the unsaturated condition, hydraulic conductivity (K) in Eq. 1.18 needs to be replaced by unsaturated conductivity (Kk_r). In this case, Eq. 1.36 becomes (Liu 2014)

$$k_{rM} = \left(1 - \frac{1 - \mathrm{e}^{-\frac{i}{I_{un}}}}{\frac{i}{I_{un}}}\right)k_r \tag{1.37}$$

Herein, we denote I and I_{un} as the threshold hydraulic gradient under the saturated and unsaturated conditions, respectively. To calculate I_{un} with Eq. 1.24, the permeability (k) there needs to be replaced by unsaturated permeability (kk_r) (Liu and Birkholzer 2013). As previously indicated, water flow under the unsaturated condition generally exhibits stronger non-Darcian flow behavior than that under the saturated condition, because the former corresponds to smaller sizes of pores occupied by water.

For demonstration purposes, and because of the lack of alternatives, we assume that k_r in Eq. 1.37 can be represented by Eq. 1.34. When hydraulic gradient (i) in a test is given, k_{rM} can be calculated from Eq. 1.37 with I_{un} and k_r obtained from Eqs. 1.24 and 1.34, respectively. The value for m is assigned to be 0.6 (Zheng et al. 2012). Figure 1.14 shows the calculated k_{rM} curves as a function of the effective saturation. For each curve, the hydraulic gradient i is fixed. The threshold gradient I is determined from Eq. 1.24 with k = 5E−20 m^2. Note that $I = 0$ corresponds to the "true" relative permeability curve calculated with Eq. 1.34.

Figure 1.14 shows that a measured relative permeability, for a given saturation, is significantly smaller than the "true" value, especially for the low saturations under which non-Darcian flow behavior becomes relatively strong. The measured curve is a function of the hydraulic gradient used in the test. A larger gradient gives results closer to the "true" values, which is expected from Fig. 1.5. If relatively small hydraulic gradients are used in tests, the measured relative permeability can be smaller than one under the saturated condition, because permeability (k) measurements are not made in the linear regime of the relationship between water flux and hydraulic gradient (Fig. 1.5). The differences between "measured" curves and the "true" curve would be even more dramatic if Eq. 1.21 with α values larger than one was used for describing the non-Darcian flow behavior, because it gives much smaller values for dq/di when the hydraulic gradient is smaller than the corresponding threshold gradient (Eq. 1.20). Nevertheless, Fig. 1.14 clearly indicates that without considering the non-Darcian flow behavior, the "measured" values for relative permeability involve large errors.

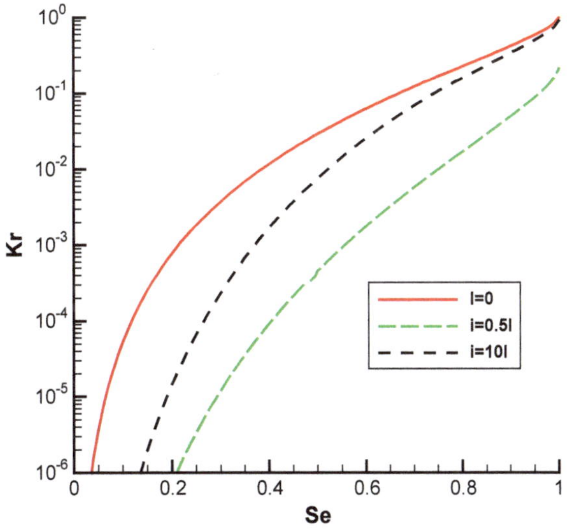

Fig. 1.14 "Measured" relative permeability curves, under different hydraulic gradients, as a function of the effective saturation (Liu 2014). Note that $I = 0$ corresponds to the "true" relative permeability curve calculated with Eq. 1.34

The finding seems to be consistent with the relative permeability measurements of rock matrix from the unsaturated zone of Yucca Mountain, Nevada, USA (BSC 2004). (More information on the Yucca Mountain Project will be given in Chap. 2.) Those measurements, for a given saturation, are dramatically smaller than what are predicted from the van Genuchten relationship (Eq. 1.34) for samples with (absolute) permeability values below 10^{-18} m^2. The differences can be explained with the non-Darcian flow behavior in low-permeability media, as discussed above.

There are several ways to recover the "true" relative permeability curve from raw measurements by incorporating the non-Darcian flow behavior (Liu 2014). The ideal one would be to use large enough hydraulic gradients such that throughout the tests, the relationship between water flux and the gradient is in the linear regime (Fig. 1.5). However, this may not be feasible in practice because the threshold gradient can be huge for low saturations. When the relationship between permeability and the threshold gradient (e.g., Eq. 1.24) is known with confidence, the only unknown in Eq. 1.37 is k_r for a given measurement k_{rM}. In this case, k_r can be estimated by solving Eqs. 1.24 (generalized for the unsaturated condition) and 1.37. If the required relationship between permeability and the threshold gradient is not known for a specific rock type, measurements for relative permeability with different hydraulic gradients, such as those shown in Fig. 1.14, need to be made. To recover the true curve, we can simultaneously fit the measured curves using Eqs. 1.24, 1.34 and 1.37 with parameters A', B', and m being treated as fitting parameters. The advantage of this approach is that both the true relative permeability curve and the relationship between permeability and threshold gradient are simultaneously determined from measurements.

1.7.3 Imbibition of Fracturing Fluids into Shale Matrix and a Methodology to Determine Relevant Parameters

Hydraulic fracturing of horizontal wells is a widely used technique for recovering natural gas from shale gas and other unconventional reservoirs characterized by an extremely low permeability and generally involves the use of large volume of fracturing fluids along with proppants. The water-based fracturing fluids (that are now most commonly used in hydraulic fracturing) include about 99 % of fresh or recycled water, complimented by the addition of chemicals like surfactants, friction reducers, biocides, clay stabilizers and scale inhibitors (e.g., Roychaudhuri et al. 2013). During hydraulic fracturing and the well shut-in stages, significant amount of fracturing fluids will flow into surrounding shale from hydraulic fractures through imbibition process. Spontaneous imbibition is of particular interest, because it is the major mechanism for liquid uptake by the formation during relatively long well shut-in time period before gas production starts (Roychaudhuri et al. 2013; Rangel-German and Kovscek 2002). The liquid imbibition could cause the loss of gas relative permeability and lead to chemically altered zone near fracture-matrix interface. Obviously, accurately predicting the imbibition process has many practical implications for shale-gas production (Liu et al. 2015).

Considerable evidence of the non-Darcian imbibition behavior exists in the shale gas literature (Liu et al. 2015). For example, several research groups experimentally demonstrated that the cumulative liquid mass of imbibition into shale samples (from spontaneous imbibition laboratory tests) as a function of time generally does not follow a straight line with a slope of 0.5 in the log-log plots (Hu and Ewing 2014; Roychaudhuri et al. 2013). The slope of 0.5 is a signature of Darcy flow, as will be discussed later. However, this non-Darcian liquid flow behavior has not been given enough attention in modeling fluid flow in shale gas reservoirs. This subsection presents a phenomenological model for unconventional liquid imbibition process and a methodology to estimate the associated model parameters from laboratory tests (Liu et al. 2015).

Section 1.3 reviews the currently available relationships between water flux and hydraulic gradient and proposes the generalized Darcy's law. The model of Liu et al. (2015) is consistent with the generalized Darcy's law and related experimental observations. This new development also allows for establishing a relatively simple laboratory procedure to determine key parameters associated with the non-Darcian liquid flow. As previously indicated, spontaneous imbibition is of particular interest, because it is the major mechanism for liquid uptake by the formation during relatively long well-shut time period (Roychaudhuri et al. 2013; Rangel-German and Kovscek 2002). Thus, this subsection is devoted to spontaneous imbibition that is driven by the capillary force. The content of this subsection is based on the materials from Liu et al. (2015).

The key assumption is that liquid flux is a power function of capillary pressure gradient, or

$$q_x = -\frac{k^*}{\mu}\left|\frac{\partial p_c}{\partial x}\right|^{n-1}\frac{\partial p_c}{\partial x} \tag{1.38}$$

where k^* is an analogue of permeability for unsaturated liquid flow (or permeability multiplied by relative permeability), p_c is capillary pressure (that is negative), n is a positive parameter, x is location, and again μ is fluid viscosity. Equation 1.38 will be reduced to Darcy's law for $n = 1$. Note that liquid flux q_x can be positive or negative to represent its direction. The assumption (Eq. 1.38) is justified by the following two considerations. First, as shown in Fig. 1.15, Eq. 1.21 can be approximately represented by a power-law function for a large range of hydraulic gradient ($i/I < 4$) that is capillary pressure gradient divided by fluid density and by gravitational acceleration. Second, more importantly, this treatment is consistent with spontaneous imbibition observations, as will be shown later.

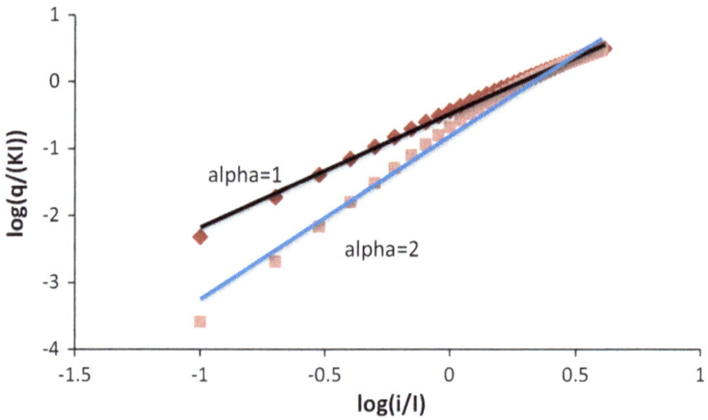

Fig. 1.15 Approximation of Eq. 1.21-1 by power functions for $i/I < 4$ (Liu et al. 2015)

During the imbibition process, a unique relationship between capillary pressure and volumetric liquid content θ exists. Thus, we have

$$\frac{\partial p_c}{\partial x} = \frac{dp_c}{d\theta}\frac{\partial \theta}{\partial x} \tag{1.39}$$

Then Eq. 1.38 can be rewritten as

$$q_x = -D(\theta)\left|\frac{\partial \theta}{\partial x}\right|^{n-1}\frac{\partial \theta}{\partial x} \tag{1.40}$$

where

$$D(\theta) = \frac{k^*}{\mu}\left|\frac{dp_c}{d\theta}\right|^n \tag{1.41}$$

Combining Eq. 1.40 with the water volume balance yields

$$\frac{\partial \theta}{\partial t} = -\frac{\partial q_x}{\partial x} = \frac{\partial}{\partial x}\left(D(\theta)\left|\frac{\partial \theta}{\partial x}\right|^{n-1}\frac{\partial \theta}{\partial x}\right) \tag{1.42}$$

where t is time. Because water density is considered to be constant, the water volume balance is equivalent to water mass balance. We have so far concentrated on one-dimensional flow case, simply because the imbibition process can be considered one dimensional. Note that penetration depth of fracturing fluids into shale matrix is generally small as a result of low shale permeability.

In order to use Eq. 1.42 to simulate liquid flow, we need to know parameter n and relationship $D(\theta)$. Liu et al. (2015) provide a solution to Eq. 1.42 that can serve as the base for experimentally determining them in the laboratory under certain conditions.

Liu et al. (2015) consider an infinite long shale column subject to the imbibition process (from the inlet $x = 0$) and the following boundary and initial conditions:

$$\theta(x,t) = \theta_i \quad (x \geq 0, t = 0) \tag{1.43-1}$$

$$\theta(x,t) = \theta_0 \quad (x = 0, t > 0) \tag{1.43-2}$$

$$\theta(x,t) = \theta_i \quad (x \to \infty, t > 0) \tag{1.43-3}$$

For the problem under consideration, the liquid content gradient is negative. Thus, Eq. 1.42 can be rewritten as

$$\frac{\partial \theta}{\partial t} = -\frac{\partial}{\partial x}\left(D(\theta)\left|\frac{\partial \theta}{\partial x}\right|^n\right) \tag{1.44}$$

Using the transformation

$$\lambda = xt^{-\frac{1}{n+1}} \tag{1.45}$$

Liu et al. (2015) transform Eqs. 1.43 and 1.44 into

$$\theta(\lambda) = \theta_i \quad (\lambda \to \infty) \tag{1.46-1}$$

$$\theta(\lambda) = \theta_0 \quad (\lambda = 0) \tag{1.46-2}$$

and

$$\frac{\lambda}{n+1}\frac{d\theta}{d\lambda} = \frac{d}{d\lambda}\left[D(\theta)\left|\frac{d\theta}{d\lambda}\right|^n\right] \tag{1.47}$$

Note that Eq. 1.47 can be much easily solved than Eq. 1.44, although they are equivalent. This is because Eq. 1.47 is an ordinary differential equation with λ as the only independent variable.

Directly integrating Eq. 1.47 for the interval (λ, ∞) yields

$$D(\theta) = \frac{\int_{\theta_i}^{\theta} \frac{\lambda}{n+1} d\theta}{\left|\frac{d\theta}{d\lambda}\right|^n} \tag{1.48}$$

It indicates that $D(\theta)$ can be calculated when $\theta(\lambda)$ is given.

Based on the liquid mass balance in the whole system, one can determine cumulative imbibition (in terms of liquid volume) as

$$M(t) = A_{cs}\int_{0}^{\infty}(\theta-\theta_i)dx = A_{cs}(\theta-\theta_i)x|_0^{\infty} - A_{cs}\int_{\theta_0}^{\theta_i}xd\theta = A_{cs}\int_{\theta_i}^{\theta_0}xd\theta \quad (1.49)$$

where A_{cs} is cross-sectional area of the shale column. Combining Eqs. (1.49) and (1.45) gives

$$M(t) = \left(A_{cs}\int_{\theta_i}^{\theta_0}\lambda d\theta\right)t^{\frac{1}{n+1}} \quad (1.50)$$

Thus, the cumulative mass is a power function of time. For Darcian flow (n = 1), the exponent is equal to 0.5. Obviously, for non-Darcian liquid flow, the exponent is not 0.5 anymore. Also note that for n = 1, the results (Eqs. 1.48 and 1.50) will be reduced to the classic results of Bruce and Klute (1956) for Darcian flow; this study can be considered a generalization of the method of Bruce and Klute (1956).

The theoretical development discussed above relies on a key assumption that liquid flux is a power function of pressure gradient (Eq. 1.38). This results in an interesting relationship between cumulative imbibition and time (Eq. 1.50). The usefulness of the assumption and related theory can be demonstrated by a comparison between that relationship and the relevant experimental observations. In this section, Eq. 1.50 is compared with data of spontaneous imbibition tests reported by Hu and Ewing (2014) and Roychaudhuri et al. (2013), because the data sets and the associated laboratory test procedures are well documented by these authors and the related test conditions are consistent with those specified in Eq. 1.43.

In Hu and Ewing (2014), all the test samples are from the Barnett shale formation. The samples were cut into rectangular prisms at about 15 mm and then oven-dried at 60 °C for at least 48 h before being subject to the imbibition experiments. The sample bottom was submerged to a depth 1 mm in a water reservoir. Care was taken of to minimize (or eliminate) evaporation from other surfaces of samples and allow air escape from the top. The cumulative imbibition was monitored by automatically recording the sample weight change over time. Observations for a typical sample are shown in Fig. 1.16. At the very early time (less than 1 min), imbibition increases abnormally quickly, which could result from the boundary effects or balance stability problem. However, after that, the relationship between cumulative imbibition and time follows a nice straight line in the log-log scale with a slope of 0.262. Obviously, this is very much consistent with Eq. 1.50 with n = 2.82. They found the similar behavior (with slopes ranging from 0.214 to 0.357) for all the other samples they investigated. This further confirms the existence of non-Darcian liquid flow behavior in shale. As previously indicated, Darcian flow behavior corresponds to a slope of 0.50.

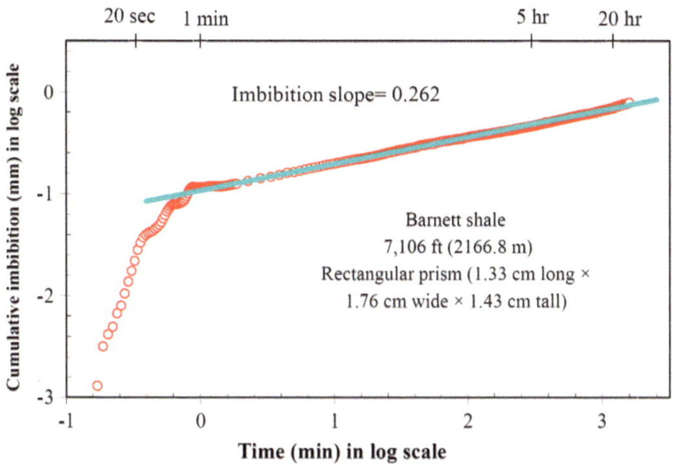

Fig. 1.16 Observed cumulative imbibition as a function of time for a Barnett shale sample (from Hu and Ewing 2014)

Hu and Ewing (2014) attributed the observed non-Darcian flow behavior to poor pore connectivity in low-permeability media. However, it is clearly demonstrated mathematically in the above derivation that the observed non-Darcian liquid flow is a direct result of the nonlinearity between imbibition flux and pressure gradient. We also believe that this nonlinearity is largely caused by the strong solid–liquid interaction (Miller and Low 1963; Hansbo 1960), rather than pore connectivity. This argument, for example, is strongly supported by the experimental results of Xu et al. (2007). As previously indicated, he experimentally investigated the relationship between flux of deionized water and pressure gradient in individual micro-tubes with diameters ranging from 2 to 30 μm, and found that water flow in micro-tubes with diameters of less than 16 μm becomes significantly nonlinear. Obviously, their results have nothing to do with pore connectivity.

The similar test procedure was used by Roychaudhuri et al. (2013) in their spontaneous imbibition tests of shale samples from the Marcellus formation. They noticed that in the plot of cumulative imbibition versus $t^{1/2}$, the observation cannot be represented by a single straight line. If they could, the liquid flow would follow Darcy's law. Instead, in that plot, they used two separated straight lines with two different slopes to represent the data, and also interpreted the first straight line (with a relatively large slope) as a signature of imbibition to micro-fractures. While their interpretation is of interest, their data can be adequately represented by Eq. 1.50 (Fig. 1.17), which is consistent with the results of Hu and Ewing (2014). Also note that Sample #11 has a slope of 0.47 in Fig. 1.17 that is very close to 0.5, indicating that water flow is close to Darcian flow in this sample. Nevertheless, most shale samples are characterized by unconventional flow and our theory (discussed above) can be applied to both Darcian and non-Darcian flow processes.

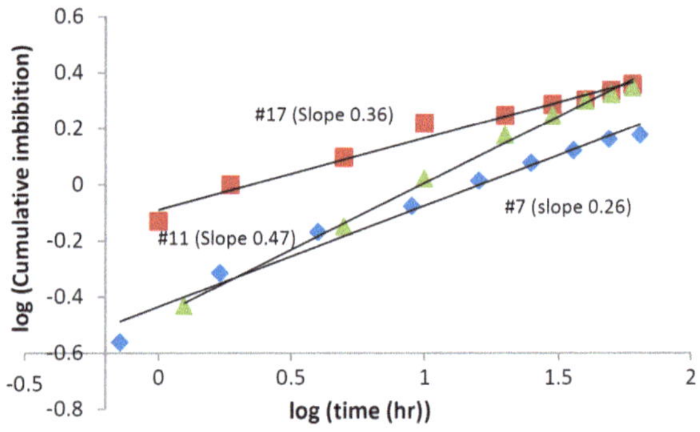

Fig. 1.17 Comparisons between the imbibition data of Roychaudhuri et al. (2013) and the theory (Eq. 1.50) (Liu et al. 2015)

The non-equilibrium imbibition has also been reported to give slope values (in the log-log plots shown in Figs. 1.16 and 1.17) different from 0.5 (Barenblatt and Gilman 1987; Silin and Patzek 2004; Guen and Kovscek 2006). The non-equilibrium theory assumes that the redistribution of the fluids in the pore space with changing saturation is not instantaneous, but takes some time, while the flow process still follows Darcy's law. Silin and Patzek (2004) provided a detailed discussion of the theory and the related mathematical development. They found that cumulative imbibed liquid volume, under the non-equilibrium condition, is described by Eq. 1.51 for early time and Eq. 1.52 for late time.

$$M(t) \propto t^{\frac{1}{2}+\frac{1}{2(1+n_0)}} \tag{1.51}$$

where n_0 is a positive constant.

$$M(t) \propto \left(1 - e^{-\frac{t}{\tau_0}}\right)\left(\frac{t}{\tau_0}\right)^{1/2} \tag{1.52}$$

where τ_0 is called relaxation time that is a positive constant as well. Obviously, Eq. 1.51 shows that at the early time the cumulative volume (or mass) is indeed a power function of time with the exponent that is different from 0.5, or more specifically larger than 0.5. (The exponent here corresponds to the slope in the log-log plots.) Because the term $1 - e^{-t/\tau_0}$ increases with time, if one uses a power function of time to approximate Eq. 1.52, the exponent would be larger than 0.5. Thus, the non-equilibrium always gives exponent larger than 0.5, which is not consistent with the theoretical and observed results for non-Darcian imbibition that corresponds to an exponent less than 0.5 (Figs. 1.16 and 1.17). In other words, the non-equilibrium is not the mechanism for non-Darcian imbibition. Our argument is consistent with findings by Schmid and Geiger (2012, 2013). Based on their analyses of a large number of spontaneous imbibition test data sets, Schmid and Geiger (2012, 2013) concluded that consideration of the non-equilibrium is not necessary to explain those data sets.

As indicated in Eq. 1.40, parameter n and function $D(\theta)$ need to be given for modeling non-Darcian liquid flow. Based on the analytical results for one-dimensional spontaneous imbibition process (Eqs. 1.48 and 1.50), Liu et al. (2015) propose a laboratory test procedure to estimate n and $D(\theta)$. The laboratory test should be designed in such a way that test conditions are consistent with initial and boundary conditions to obtain Eqs. 1.48 and 1.50. The imbibition occurs from one end of a shale column (not from its sides) and the length of the column should be long enough such that it can be approximately considered infinite for the imbibition. The latter should be easily satisfied in practice because imbibition into shale sample is a slow process. Care also needs to be taken such that evaporation from the shale column is eliminated or minimized. Then using a procedure similar to that of Hu and Ewing (2014), the cumulative imbibition as a function of time, $M(t)$, is monitored. The log-log plot of $M(t)$, like Figs. 1.16 and 1.17, allows for determination of parameter n by fitting the data with Eq. 1.50; the slope in the plot is equal to $1/(n + 1)$.

To estimate $D(\theta)$ from Eq. 1.48, we need to experimentally determine $\theta(\lambda)$ (or $\lambda(\theta)$). Based on the transformation given in Eq. 1.45, $\theta(\lambda)$ can be calculated from θ spatial distribution for a given time, or from θ as a function of time at a given location, when parameter n is known. In this study, the use of θ spatial distribution at the time when imbibition test is finished is considered more practical. Since the imbibition depth is likely small, a high-resolution measurement of θ distribution along the shale column is required.

For many practical reasons, an observed θ spatial distribution is often not smooth and involves certain degree of fluctuation, which may cause some problem with calculating $D(\theta)$ from Eq. 1.48 when directly observed $\theta(\lambda)$(or $\lambda(\theta)$) is employed. This issue has been addressed by many studies on applications of the method of Bruce and Klute (1956) to determining soil water diffusivity. Since θ distributions generally follow certain regular shapes, a number of empirical algebraic expressions for $\theta(\lambda)$ exist and can be used to fit observed $\theta(\lambda)$ data adequately (e.g., Brutsaert 1982; Evangelides et al. 2010); these expressions correspond to smooth curves. Furthermore, these expressions allow for derivation of algebraic expressions for $D(\theta)$ that are convenient to use in modeling studies.

To demonstrate the usefulness of the treatment discussed above, we focus on a relatively simple expression for $\lambda(\theta)$:

$$\lambda = \beta(1 - \theta^{*^{\alpha'}}) \tag{1.53}$$

where

$$\theta^* = \frac{\theta}{\theta_0} \tag{1.54}$$

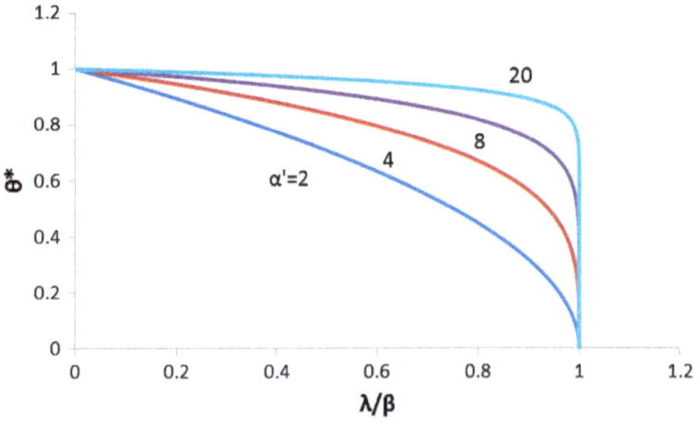

Fig. 1.18 Curves calculated from Eq. 1.53 for several α′ values (Liu et al. 2015)

Figure 1.18 shows $\theta^*(\lambda)$ curves for several α' values. A large α' value corresponds to a relatively sharp liquid content profile and a small α' value to a gradually varying profile. Inserting Eq. 1.53 into Eq. 1.48 yields (Liu et al. 2015)

$$D(\theta) = \frac{\beta^{n+1}(\alpha' - 1)^n}{\theta_0^{n-1}} \left[(\theta^* - \theta_i^*) - \frac{(\theta^*)^{\alpha'+1} - (\theta_i^*)^{\alpha'+1}}{\alpha' + 1} \right] (\theta^*)^{(\alpha'-1)n} \tag{1.55}$$

Values for parameters α' and β can be determined by fitting the $\theta(\lambda)$ data with Eq. 1.53, which is subject to the constraint:

$$\int_{\theta_i}^{\theta_0} \lambda d\theta = \beta\theta_0 \left[(1 - \theta_i^*) - \frac{1 - (\theta_i^*)^{\alpha'+1}}{1 + \alpha'} \right] \tag{1.56}$$

where the value for the left hand side of Eq. 1.56 is determined when the cumulative imbibition ($M(t)$) data is fitted with Eq. 1.50. From the derivation of Eq. 1.50, it is evident that Eq. 1.56 is a constraint associated with liquid mass (volume) balance.

1.7.4 Non-Darcian Flow and Abnormal Liquid Pressure in Shale Formations

The non-Darcian flow behavior is consistent with the often observed abnormal liquid pressure in shale formations. The abnormal pressure refers to the pore pressure that is considerably higher (overpressure) or lower (underpressure) than the hydrostatic pressure (Deming 1994; Tremosa et al. 2012). The hydrostatic pressure is defined as the pore pressure when fluid column is in equilibrium with gravity under the condition that pores are well connected in the vertical direction.

It is also well known that a shale formation is capable of confining anomalous pressure over geologic time, or plays a role as a pressure seal. The existence of abnormal pressure within shales themselves and the capability of shales to confine pressure are closely related because both of them result from the same fact that shale formations can prevent considerable pressure propagation. Based on his analyses of pressure propagation processes, Deming (1994) concluded that the permeability needed for a geologic unit to act as a pressure seal over a time span of about one million years is about 10^{-21} to 10^{-23} m^2, a range generally lower than most measurements of shale permeability. Note that the use of lower permeability values than actual ones is essentially equivalent to the consideration of non-Darcian flow behavior when the hydraulic gradient is less than the threshold gradient (Fig. 1.5). The dominant mechanism for the anomalous pressure in shale formations has been an issue of debate in the literature. The osmotic pressure within shale formations has also been proposed as a major mechanism (e.g., Tremosa et al. 2012). It may contribute to the persistence of observed overpressures, but cannot explain the frequently observed pore pressures that are lower than hydrostatic ones in shale formations.

Figure 1.19 demonstrates the impact of non-Darcian flow on the existence of abnormal pressures within shale formations. Again for simplicity, we use Eq. 1.18 for describing the non-Darcian flow behavior. Consider that initially there is a pressure discontinuity at interfaces between a shale formation and upper and lower adjacent formations, which leads to infinite pressure (hydraulic) gradient there. As indicated by Eq. 1.18, shale-formation water near the interfaces is mobile for the given gradient and released into the adjacent formations. Consequently, pressure gradients near the interfaces decrease with time until hydraulic gradient i reaches its threshold value I. Based on Eq. 1.18, water is not mobile anymore below the hydraulic gradient less than I. Thus, a pressure distribution (the bolder curve in Fig. 1.19) will not change anymore with time, leading to a steady-state abnormal pressure distribution.

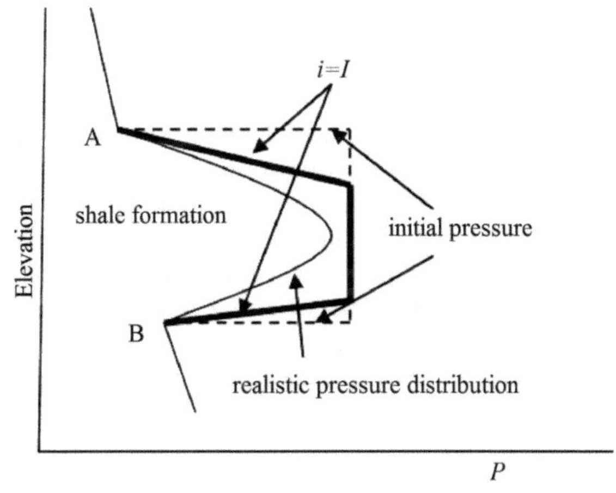

Fig. 1.19 Demonstration of the impact of non-Darcian flow on the pressure seal effect. P refers to pore water pressure. Elevations at A and B correspond to interfaces between shale formation and the overlying and underlying formations, respectively

While the process for the formation of abnormal pressure discussed above is, in principle, applicable to a realistic case, two issues need to be taken into consideration for determining a more realistic pressure distribution in a shale formation. Firstly, non-Darcian flow is actually described by our generalized Darcy's law (Eq. 1.21), rather than Eq. 1.18. Based on Eq. 1.21, liquid water will never be completely immobile except for a zero hydraulic gradient. Thus, a real pressure distribution should be smoother than that predicted with Eq. 1.18 (the bolder line in Fig. 1.19). Secondly, since water is always mobile for non-zero pressure gradient, the real pressure distribution is transient and not in a steady state. However, because water flow is extremely slow for relatively low hydraulic gradient (Fig. 1.5), a shale formation can contain or be capable of confining anomalous pressure over geologic time. A more realistic pressure distribution within a shale formation is also given in Fig. 1.19 as well.

1.8 Concluding Remarks

(A) The key result from this chapter is the generalized Darcy's law given as

$$q = K\left[i - \frac{I}{\gamma\left(\frac{1}{\alpha}\right)}\gamma\left(\frac{1}{\alpha}, \left(\frac{i}{I^*}\right)^{\alpha}\right)\right] \tag{1.21}$$

where the threshold hydraulic gradient I and parameter α characterize impacts of solid–liquid interaction and pore size distribution, respectively. The above relationship is very general in a sense that it includes Darcy's law and the two commonly used forms of modified Darcy's law as its special cases. The consistency between the above equation and experimental observations from different sources is demonstrated. Extensions of the above equation from one-dimensional to multidimensional cases are explored.

(B) The generalized Darcy's law and its variations are used to attack several key technical issues facing the geoscience community, including the relative importance of diffusion in the excavation damaged zone (EDZ) for a shale repository of high-level nuclear waste, the accurate measurement of relative permeability for multiphase flow in a low-permeability porous medium, non-Darcian flow behavior during imbibition of fracturing fluids into a shale gas reservoir, and formation of the pressure seal in shale formations.

(C) Our generalized Darcy's law (Eq. 1.21) is phenomenological in nature. Its mathematical form may be refined by more fundamental studies on the impact of solid–liquid interaction on liquid flow. Furthermore, influences of environmental factors (e.g., temperature and pressure) on the threshold hydraulic gradient I and parameter α are not totally clear yet.

References

Babour SL, Fredlund DG, Pufah DE (1991) The osmotic role in the behaviour of swelling clay soils. Proceedings of the NATO advanced research workshops, Korfu, Greece, 1–6 July 1991

Barenblatt G, Gilman A (1987) A methmatical model of non-eqilibrium countercurrent capillary imbibition. Eng Phys J 52(3):46–461

Bear J (1979) Hydraulics of groundwater. McGraw-Hill Inc., New York

Bianchi M, Liu HH, Birkholzer J (2013) Radionuclide transport behavior in a generic radioactive waste repository. Groundwater. doi:10.1111/gwat.12171

Blecker RF (1970) Saturated flow of water through clay loam subsoil material of the Brolliat and Springerville soil series. Master Thesis, The University of Arizona

Brown GO (2002) Henry Darcy and the making of a law. Water Resour Res. doi:10.1029/2001WR000727

Bruce RR, Klute A (1956) The measurement of soil moisture diffusivity. Soil Sci Soc Am Proc 20:458–462

Brutsaert W (1982) Some exact solutions for nonlinear desorptive diffusion. J Appl Math Phys 33:540–546

BSC (2004) Analysis of hydrologic properties data. Report ANL-NBS-HS-000042, Yucca Mountain Project

Chen X, Cao GX, Han AJ et al (2008) Nanoscale fluid transport: size and rate effects. Nano Lett 8 (9):2988–2992

Cui YJ, Tang AM, Loiseau C et al (2008) Determining the unsaturated hydraulic conductivity of a compacted sand-bentonite mixture under constant-volume and free-swell conditions. Phys Chem Earth 33:S462–S471

Daniel H (1982) Introduction to soil physics. Academic Press, San Diego, CA

Darcy H (1856) The public fountains of the city of Dijon. Dalmont, Paris

Deming D (1994) Factors necessary to define a pressure seal. AAPG Bulletin 78(6):1005–1009

Dubin B, Moulin G (1986) Influences of critical gradient on the consolidation of clay. In: Yong RN, Townsend FC (eds) Consolidation of soils: testing and evaluation. West Conshohocken, PA

Dupuit AJEJ (1857) Essay on movement of water through permeable terrains. C R Hebdomadair Seanc Acad Sci (Paris) 45:92–96

Evangelides C, Arampatzis G, Tzimopoulos C (2010) Estimation of soil moisture profile and diffusivity using simple laboratory procedures. Soil Sci 175(3):118–127

Farrow MR, Chremos A, Camp PJ et al (2011) Molecular simulations of kinetic-friction modification in nanoscale fluid layers. Tribol Lett 42:325–337

Freeze RA (1994) Henry Darcy and the fountains of Dijon. Groundwater 32(1):23–30

Guen SSL, Kovscek AR (2006) Nonequilibrium effects during spontaneous imbibition. Transp Porous Media 63:127–146

Hansbo S (1960) Consolidation of clay with special reference to influence of vertical sand drains, Swed Geotech Inst Proc 18, Stockholm

Hansbo S (2001) Consolidation equation valid for both Darcian and non-Darcian flow. Geotechnique 51(1):51–54

He S, Liu H, Qin G (2015) Molecular dynamics simulation on modeling shale gas transport and storage mechanisms in complex nano-pore structure in organic matters. Paper SPE 178713 presented in the unconventional resources technology conference held in San Antonio, Texas, USA, 20–22 July 2015

Hu QH, Ewing RP (2014) Integrated experimental and modeling approaches to studying the fracture-matrix interaction in gas recovery from Barnett shale. Report 09122-12, University of Texas at Arlington (Prepared for US Department of Energy)

Hubbert MK (1940) The theory of groundwater motion. J Geol 48(8):758–944
Kang JB (2008) Membrane behaviour of clay liners. PhD Thesis, Colorado State University
Liu HH (2014) Non-Darcian flow in low-permeability porous media: key issues related to geological disposal of high-level nuclear waste in shale formations. Hydrogeol J 22(7):1525–1534
Liu HH, Birkholzer J (2013) On the relationship between water flux and hydraulic gradient for unsaturated and saturated clay. J Hydrol 476:242–247
Liu HH, Lai BT, Chen JH (2015) Unconventional spontaneous imbibition into shale matrix: theory and a methodology to determine relevant parameters. Transp Porous Med. doi:10.1007/s11242-015-0580-z
Liu HH, Li LC, Birkholzer J (2012) Unsaturated properties for non-Darcian water flow in clay. J Hydrol 430–431:173–178
Liu HH, Zhang YQ, Zhou Q et al (2007) An interpretation of potential scale dependence of the effective matrix diffusion coefficient. J Contam Hydrol 90(1–2):41–57
Lutz JF, Kemper WD (1959) Intrinsic permeability of clay as effected by clay-water interaction. Soil Sci 88:83–90
Ma M, Shen L, Sheridan J, Liu Z et al (2010) Friction law for water flowing in carbon nanotubes. In: 2010 international conference on nanoscience and nanotechnology, Sydney, Australia, 20–22 Feb 2010
Miller RJ, Low PF (1963) Threshold gradient for water flow in clay systems. Soil Sci Soc Am Proc 27(6):605–609
Narasimhan TN (2005) Hydrogeology in North America: past and future. Hydrogeol J 13:7–24
Philip JR (1995) Desperately seeking Darcy in Dijon. Soil Sci Soc Am J 59:319–324
Rangel-German ER, Kovscek AR (2002) Experimental and analytical study of multi-dimensional imbibition in fractured porous media. J Petrol Sci Eng 36(1–2):45–60
Roychaudhuri R, Tsotsis TT, Jessen K (2013) An experimental investigation of spontaneous imbibition in gas shales. J Petrol Sci Eng 111:87–97
Schmid KS, Geiger S (2012) Universal scaling of spontaneous imbibition for water-wet systems. Water Resour Res. doi:10.1029/2011WR011566
Schmid KS, Geiger S (2013) Universal scaling of spontaneous imbibition for arbitrary petrophysical properties: water-wet and mixed-wet states and Handy's conjecture. J Petrol Sci Eng 101:44–61
Silin D, Patzek T (2004) On Barenblatt's model of spontaneous countercurrent imbibition. Transp Porous Media 54:297–322
Simmons CT (2008) Henry Darcy (1803–1858): immortalised by his scientific legacy. Hydrogeol J 16:1023–1038
Swartzendruber D (1961) Modification of Darcy's law for the flow of water in soils. Soil Sci 93:22–29
Tremosa J, Gonçalvès J, Matray JM (2012) Natural conditions for more limited osmotic abnormal fluid pressures in sedimentary basins. Water Resour Res. doi:10.1029/2011WR010914
Tsang CF, Barnichon JD, Birkholzer J et al (2012) Coupled thermo-hydro-mechanical processes in the near field of a high-level radioactive waste repository in clay formations. Int J Rock Mech Min Sci 49:31–44
van Genuchten M (1980) A closed-form equation for predicting the hydraulic conductivity of unsaturated soil. Soil Sci Soc Am J 44(5):892–898
Wang XX, Yang ZM, Sun YP et al (2011) Experimental and theoretical investigation of nonlinear flow in low permeability reservoir. Procedia Environ Sci 11:1392–1399
Xu SL, Yue XA, Hou JR (2007) Experimental investigation on flow characteristics of deionized water in microtubes. Chin Sci Bull 52(6):849–854
Zheng L, Li L, Rutqvist J et al (2012) Modeling radionuclide transport in clays. Report FCRD-URD-2012-000128, Lawrence Berkeley National Laboratory

Zhou QL, Birkholzer JT, Tsang CF (2008) A method for quick assessment of CO_2 storage capacity in closed and semi-closed saline formations. Int J Greenh Gas Control 2(4):626–639

Zou Y (1996) A nonlinear permeability relation depending on the activation energy of pore liquid. Geotechnique 46(4):769–774

Chapter 2
Generalization of the Darcy-Buckingham Law: Optimality and Water Flow in Unsaturated Media

Abstract The Darcy-Buckingham law extends Darcy's law from single phase flow in the subsurface to multiphase flow and was discovered by Edgar Buckingham (1867–1940). It is valid under the local-equilibrium condition that, however, does not always hold especially when fingering flow occurs. This chapter is devoted to generalizing the Darcy-Buckingham law by relaxing the local-equilibrium condition. The new development is based on an optimality principle that water flow in unsaturated media self-organizes in such a way that the resistance to water flow is minimized. Unlike the traditional form of the Darcy-Buckingham law, the relative permeability in the generalization is also a function of water flux, which is a direct result of the non-equilibrium flow behavior. The generalized Darcy-Buckingham law has been shown to be consistent with laboratory observations and field data for gravitational fingering flow in unsaturated soils. It is also the theoretical foundation for the active fracture model, the key constitutive relationship for modeling flow and transport in the unsaturated zone of Yucca Mountain, Nevada, that is the national geological disposal site for high-level nuclear waste in USA.

Darcy's law has been regarded as the most fundamental physical law in describing fluid flow in the subsurface. This law, however, should be called the Darcy-Buckingham law for multiphase flow, because Edgar Buckingham (1867–1940), an American physicist, independently discovered the relationship between water flux and hydraulic gradient in a water-air two-phase flow system (or unsaturated soil); Darcy's law was initially discovered for single phase flow only. This chapter discusses the Darcy-Buckingham law, its valid range, and our work on extending this important physical law based on an optimality principle that the water flow pattern in unsaturated media is self-organized in such a way that total water flow resistance is minimized. The developments in this chapter are mainly based on materials from Liu et al. (1998, 2011a, b), Sheng et al. (2009) and Liu (2010).

H.-H. Liu, *Fluid Flow in the Subsurface*, Theory and Applications of Transport in Porous Media 28, DOI 10.1007/978-3-319-43449-0_2

2.1 Edgar Buckingham and His Law for Water Flow in Unsaturated Soils

Largely because Edgar Buckingham (Fig. 2.1) had worked on water flow in unsaturated soils for a relatively short time period, his original contributions to multiphase flow theory are not well known beyond the soil physics community, although the Darcy-Buckingham law has been used widely in many areas including reservoir engineering. Nimmo and Landa (2005) presented an excellent review of Edgar Buckingham's life and related scientific contributions. This section is based on the materials reported in Nimmo and Landa (2005).

Buckingham was born in Philadelphia, PA, in 1867. He earned a bachelor's degree in physics from Harvard University in 1887. Then he traveled to Europe and became a graduate assistant in the physics department at the University of Strasbourg and the University of Leipzig, where he studied under chemist Wilhelm Ostwald who won the Nobel Prize in 1909. At that time, the best universities in the world were all in Europe. Buckingham got his Ph.D. degree in physics from Leipzig in 1893 and then began teaching physical chemistry and physics at Bryn Mawr

Fig. 2.1 Edgar Buckingham (1867–1940) (https://en.wikipedia.org/wiki/Edgar_Buckingham)

College where he wrote a textbook on thermodynamics. He left Bryn Mawr College in 1899 as an associate professor. After a one-year stint at University of Wisconsin as an instructor in physics, he joined the US Department of Agriculture (USDA) Bureau of Soils (BOS) in 1902 as a soil physicist to work on the dynamics of gas and water flow in soils. In the BOS, he made admirable contributions to several areas. He was the first person to introduce an energy concept into multiphase fluid flow in porous media, to measure water capillary pressure as a function of water saturation in soils, and to investigate the relative importance of diffusion versus advection for gas transport in porous media. Most importantly, he discovered the Darcy-Buckingham law and documented it in his famous USDA report (Buckingham 1907). He left BOS in 1906 for the National Bureau of Standards (NBS), where he remained until retirement in 1937. His work at the NBS includes research on helium production for the military, technical oversight of NBS support for rocketry studies by Robert Goddard, and lecturing on thermodynamics at Naval Postgraduate School in Annapolis, Maryland, USA. In 1914, in the field of dimensional analysis, he published the famous π theorem, the other celebrated achievement in his career that has been used widely in many different areas (Buckingham 1914). He died in Washington DC, USA, in 1940.

The Darcy-Buckingham law was initially written as (Nimmo and Landa 2005):

$$Q = \lambda(\theta)\frac{\partial \psi}{\partial x} \tag{2.1}$$

where Q is the magnitude of water flux, θ is volumetric water content, λ is "the capillary conductivity of the soil" that is called unsaturated conductivity (closely related to relative permeability) in the modern literature, ψ is the matrix potential defined by Buckingham (1907) as "a quantity which measures the attraction of the soil at any given point for water," and x is the location coordinate in the horizontal direction. The matrix potential, a concept originated by Buckingham (1907), is closely related to capillary pressure. (Note that the symbol λ represents two different variables in Eqs. 1.45 and 2.1, respectively.)

"Buckingham's desire to measure retention curves was constrained, but not prevented, by the lack of a direct method of measuring ψ" (Nimmo and Landa 2005). The water retention curve refers to the matrix potential (or capillary pressure) as a function of water content, or $\psi(\theta)$. Based on the principle that for a vertical soil column under equilibrium conditions, the ψ value can be known from the vertical location with respect to a water table, Buckingham was able to measure retention curves (by determining vertical water-content distribution along a soil column) for the first time for several soils (Buckingham 1907). (It can be shown that under equilibrium conditions the magnitude of the capillary pressure at a location along a vertical soil column is equal to the height difference [between that location and water table] multiplied by the water density and by gravitational acceleration.) Buckingham (1907) also discussed the dependence of unsaturated conductivity on water content, including how thin-film and filled-pore flow paths

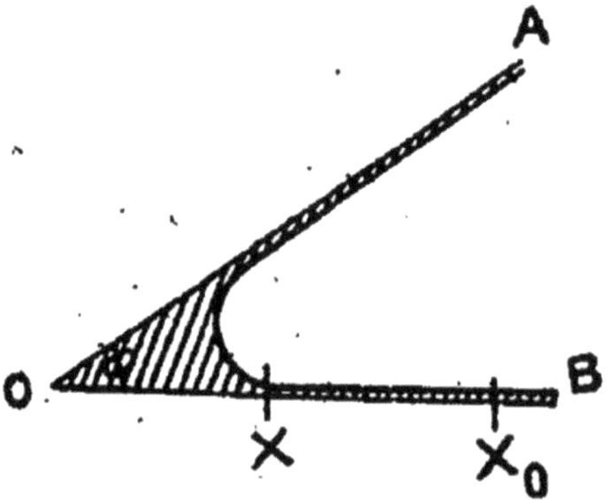

Fig. 2.2 Capillary water in a prismatic wedge (Buckingham 1907)

conduct water and how prismatic wedges near soil-particle contacts hold capillary water (Fig. 2.2).

The Darcy-Buckingham law was presented by Buckingham (1907) as a "formal analogy with Fourier's and Ohm's laws". He, however, did not mention Darcy's law in his report. While the Darcy-Buckingham law is indeed an extension or generalization of Darcy's law, it is an issue of debate as to whether Buckingham knew of Darcy's law when he did his work on water flow in unsaturated soils. The author intends to believe that he did not for the following reasons. Buckingham did not have any personal motivation to purposely ignore Darcy's law. After leaving the BOS, he never revisited his theory of multiphase flow and even never worked on flow in porous media in general (Nimmo and Landa 2005). It is very likely that Buckingham at that time did not regard his work on water flow in unsaturated soils to be a very important scientific contribution. This point of view is supported by the fact that in his report (Buckingham 1907), he even did not care to write a concluding statement in the section documenting the Darcy-Buckingham law because he wanted to rush for his new job in the NBS. It is common that scientists intend to revisit and promote their scientific works that are considered important contributions by themselves. Also, Buckingham was described as an "uncompromisingly truthful" person (Nimmo and Landa 2005).

Twenty-four years after Buckingham (1907) published his work on soil water flow, Richards (1931) combined the Darcy-Buckingham law and the continuity (mass balance) equation for water flow in his Ph.D. study and published the final result as a partial differential equation that explicitly includes gravitational potential in addition to Buckingham's matrix potential (Richards 1931). Since then, Richards' equation has been used widely in modeling unsaturated water flow in the soil science and hydrogeology communities. Five years later, Muskat and Meres (1936) and Wyckoff and Botset (1936) published their work on multiphase flow theory and the related laboratory studies within the context of reservoir engineering. In their papers, they called multiphase fluids as "heterogeneous fluids". Their theory is essentially the same as the Darcy-Buckingham law, but they did not mention Buckingham's work and gave all the credit to Henry Darcy. They were likely not aware of Buckingham's work at that time. Nevertheless, because of his original contributions to the subject, Edgar Buckingham should be regarded as the father of multiphase flow in porous media.

2.2 Unsaturated Flow Constitutive Models Under Local Equilibrium

A key assumption of the Darcy-Buckingham law is local equilibrium or uniform capillary pressure in a control volume, as evidenced in the original development procedure of the law documented in Buckingham (1907). The so-called control volume refers to a gridblock for numerical modeling, or a core sample for measuring flow properties in the laboratory. The assumption explains why the soil water conductivity in Eq. 2.1 is a function of water content (or capillary pressure) only, and has nothing to do with water flow dynamics (such as water flux), which will be further explored in this chapter. The local equilibrium assumption has been subsequently employed in many studies on determining multiphase flow properties and modeling multiphase flow processes in the subsurface. This section reviews some commonly used unsaturated flow property models, or constitutive models, to demonstrate that these models are all based on the local equilibrium assumption. For simplicity, in this and other relevant sections of this chapter, we focus on water-air flow process in soils; mechanisms and results discussed here, in principle, can be extended to other multiphase flow systems. Note that for a water-air flow process in a soil, air pressure is generally considered to be constant and equal to the atmosphere pressure. This is because air viscosity is significantly smaller than that of water, and soil air is well connected to the atmosphere. Under such conditions, soil air remains in equilibrium with the atmosphere to a good approximation.

2.2.1 Burdine Model for Relative Permeability and the Brooks-Corey Relation

Burdine (1952) developed one of the first models for relative permeability that is defined as fluid conductivity (for a phase at a given phase content) divided by the conductivity under saturated conditions. In the Burdine (1952) model, a porous medium is conceptualized as a group of individual capillary tubes each with a uniform radius corresponding to a particular pore size. Under unsaturated conditions when both water and air exist in the pores of a porous medium, all the capillary tubes in the bundle have the same capillary pressure for water because of the local equilibrium assumption.

From the Young-Laplace equation, a single capillary-tube of radius r is related to the magnitude of capillary pressure (p_c) as

$$r = \frac{2T_s \cos \beta'}{|p_c|} \tag{2.2}$$

where T_s is surface tension between water and air and β' is the contact angle. For a given water-air flow system, both T_s and β' are treated as constants. The capillary-tube radius calculated from Eq. 2.2 is called the critical radius herein. For a capillary tube with a radius equal to or less than the critical radius, it is filled with the wetting fluid (water). Otherwise, it would be filled with non-wetting fluid (air).

The water distribution in a soil is generally much more complex than what is described by a group of capillary tubes. Thus, we focus on the portion of water that is mobile and associated with capillary flow. This portion of water corresponds to an effective water content defined by $\theta_e = \theta - \theta_r$, where θ and θ_r are the actual and residual volumetric water contents, respectively. The residual water content corresponds to the portion of soil water that is immobile in a practical sense (e.g., bounding water on solid surface and water occupying dead-end pores).

Consider a soil slab isolated from a homogeneous soil column (along x direction) by two parallel cross sections normal to the x axis and separated by a distance of dx. For a given water content (or capillary pressure), the contribution of the capillary tubes with a radius r to water flux, based on Eq. 1.10, is given by

$$dq \propto -\frac{dH}{dx} r^2 d\theta_e \tag{2.3}$$

where q is water flux through the soil slab and H is again the hydraulic head defined by

$$H = z + \frac{p_c}{\rho g} \tag{2.4}$$

where z is elevation and ρ and g are water density and gravitational acceleration, respectively. Note that in Eq. 2.3, $d\theta_e$ is the fraction of the cross-sectional area corresponding to the capillary tubes with the radius r.

Integrating Eq. 2.3 and using Eq. 2.2, we have

$$q(\theta_e) \propto -\frac{dH}{dx} \int\limits_0^{\theta_e} r^2 d\theta_e \propto -\frac{dH}{dx} \int\limits_0^{\theta_e} \frac{d\theta_e}{p_c^2} \tag{2.5}$$

Based on the Darcy-Buckingham law, unsaturated hydraulic conductivity, K_{un}, for water is equal to water flux divided by hydraulic gradient. Thus, the unsaturated conductivity is obtained from Eq. 2.5 as

$$K_{un}(\theta_e) \propto \int_0^{\theta_e} \frac{d\theta_e}{p_c^2} \tag{2.6}$$

where subscript *un* refers to the unsaturated condition. Note that in Eq. 2.6, unlike in Eq. 2.1, we do not use λ to denote hydraulic conductivity to be consistent with the current literature. Equation 2.6 also implies that capillary tubes with radii larger than the critical radius do not contribute to the water conductivity because they are filled with air, not water.

Sometimes, it is more convenient to use effective saturation defined by

$$S_e = \frac{\theta - \theta_r}{\theta_{sat} - \theta_r} \tag{2.7}$$

where θ_{sat} is saturated water content. In terms of S_e, Eq. 2.6 is rewritten as

$$K_{un}(S_e) \propto \int_0^{S_e} \frac{dS_e}{p_c^2} \tag{2.8}$$

By definition, the relative permeability can be calculated by

$$k_r(S_e) = \frac{K_{un}(S_e)}{K} = \frac{\int_0^{S_e} \frac{dS_e}{p_c^2}}{\int_0^1 \frac{dS_e}{p_c^2}} \tag{2.9}$$

where K is hydraulic conductivity under the water-saturated condition. However, the above formulation is obtained under a condition that all the capillary tubes are straight and along the x axis. This is not true in reality. The water flow paths are generally tortuous in a porous medium. To take this into consideration, the right hand side of Eq. 2.9 is multiplied by an additional tortuosity factor that is an empirical parameter and generally treated as a power function of saturation. In the Burdine (1952) model, the exponent of the power function is two. Therefore, the Burdine (1952) model is given as

$$k_r(S_e) = S_e^2 \frac{\int_0^{S_e} \frac{dS_e}{p_c^2}}{\int_0^1 \frac{dS_e}{p_c^2}} \tag{2.10}$$

Equation 2.10 is for the wetting phase. For the non-wetting phase, replacing S_e with $1 - S_e$, a similar procedure can be used to obtain the non-wetting phase relative permeability as

$$k_{r,nw}(S_e) = (1 - S_e)^2 \frac{\int_{S_e}^{1} \frac{dS_e}{p_c^2}}{\int_0^1 \frac{dS_e}{p_c^2}} \tag{2.11}$$

Note that in the above equation, $(1 - S_e)$ is equal to the non-wetting phase effective saturation and the integral interval for the numerator reflects the fact that nonwetting phase occupies capillary tubes with radii larger than the critical radius.

The Burdine model and other similar models are useful because they provide correlations between relative permeability and capillary pressure (as a function of saturation). The latter can be much more easily measured than the former. Once we know $p_c(S_e)$, the relative permeability can be estimated from Eqs. 2.10 and 2.11.

There are different functional forms of $p_c(S_e)$ in the literature. One of the most commonly used forms was provided by Brooks and Corey (1964). Based on laboratory test results for a number of soil samples, they found that the relationship between water saturation and capillary pressure can be represented by

$$\begin{aligned} S_e &= \left(\frac{p_b}{p_c}\right)^{\lambda^*} && \text{for } p_c \geq p_b \\ S_e &= 1 && \text{for } p_c < p_b \end{aligned} \tag{2.12}$$

where λ^* is a parameter characterizing the pore-size distribution and p_b, called air-entry capillary pressure, is a measure of the maximum pore size forming a continuous network of flow channels within the medium (Brooks and Corey 1964).

Inserting Eq. 2.12 into Eq. 2.10 yields

$$k_r = (S_e)^{\frac{2+3\lambda^*}{\lambda^*}} \tag{2.13}$$

Similarly, combining Eqs. 2.12 and 2.11 gives relative permeability for the non-wetting phase:

$$k_{r,nw} = (1 - S_e)^2 \left(1 - S_e^{\frac{2+\lambda^*}{\lambda^*}}\right) \tag{2.14}$$

where subscript *nw* refers to the non-wetting phase.

2.2.2 Mualem Model for Relative Permeability and the van Genuchten Relation

In the Burdine (1952) model, water flow is characterized by a group of capillary tubes each of which has a uniform radius. This is not true in reality. A flow path likely consists of pores with different sizes. In other words, a flow path should be better conceptualized as a capillary tube with variable radius. To partially address this issue, Mualem (1976) developed a new conceptual model for liquid water flow in the unsaturated soil.

The water content can be related to pore size based on the fact that the contribution of water-filled pores with radii between r and r + dr to the water content is

$$d\theta_e = f(r)dr \tag{2.15}$$

where f(r) is the volumetric pore-size density distribution function.

Again consider a soil slab isolated from a homogeneous soil column (along x direction) by the two parallel cross sections. A capillary tube (flow path) with variable radius intersects the two sections where the radii are r and ξ, respectively. The water flux through capillary tubes with the same radius distribution as the one mentioned above is given by

$$dq \propto -\frac{dH}{dx} r_e^2 a_e(r,\xi)drd\xi \tag{2.16}$$

where r_e is the effective radius for the capillary tube and approximated by

$$r_e = r\xi \tag{2.17}$$

The above equation was derived for two connected pores whose radii are proportional to their lengths (Mualem 1976). Obviously, it is only an approximation considering the fact that a flow path (or capillary tube) between the two cross-sections of a soil slab consists of a number of pores with different pore radii, not necessarily two of them. The term of $a_e(r,\xi)drd\xi$ in Eq. 2.16 represents "effective area" for the corresponding capillary tubes and equals to the joint probability for r and ξ to simultaneously occur at the cross-sections of the soil slab. If sizes of pores are not spatially correlated, one has (Mualem 1976; Mualem and Dagan 1978):

$$a_e(r,\xi)drd\xi = f(r)drf(\xi)d\xi \tag{2.18}$$

Inserting Eqs. 2.17 and 2.18 into Eq. 2.16 and integrating it yields

$$\begin{aligned} q \propto -\frac{dH}{dx} \iint\limits_{\substack{warter\\ filled\\ pore}} r\xi f(r)f(\xi)drd\xi &= -\frac{dH}{dx}\int_0^r rf(r)dr\int_0^r \xi f(\xi)d\xi \\ &= -\frac{dH}{dx}\left[\int_0^r rf(r)dr\right]^2 = -\frac{dH}{dx}\left[\int_0^{\theta_e} rd\theta_e\right]^2 \end{aligned} \tag{2.19}$$

Again based on the Darcy-Buckingham law, the unsaturated water conductivity is derived from Eq. 2.19 as follows

$$K_{un}(\theta_e) \propto \left[\int_0^{\theta_e} \frac{d\theta_e}{p_c} \right]^2 \tag{2.20}$$

Following the same procedure to derive Eq. 2.9, we get

$$k_r(S_e) = \left[\frac{\int_0^{S_e} \frac{dS_e}{p_c}}{\int_0^1 \frac{dS_e}{p_c}} \right]^2 \tag{2.21}$$

Incorporating the tortuosity factor into Eq. 2.21 yields the Mualem (1976) model as

$$k_r(S_e) = S_e^{1/2} \left[\frac{\int_0^{S_e} \frac{dS_e}{p_c}}{\int_0^1 \frac{dS_e}{p_c}} \right]^2 \tag{2.22}$$

Note that the exponent value of the tortuosity factor in Eq. 2.22 is different from that in Eq. 2.10 because of different assumptions regarding the capillary tube. A procedure similar to that used to derive Eq. 2.11 can be followed to obtain the non-wetting phase relative permeability for the Mualem (1976) model. We leave the derivation to readers.

van Genuchten (1980) proposed an empirical relationship between capillary pressure head ($h = |p_c|/(\rho g)$) and effective water saturation S_e:

$$S_e = \left[\frac{1}{1 + (\alpha_{VG} h)^{n_{VG}}} \right]^m \tag{2.23}$$

where m, n_{VG}, and α_{VG} are fitting parameters. The first two parameters (m and n_{VG}) are measures of pore-size distribution and the last one is a measure of the largest pore size or the air-entry capillary pressure head. Compared with Eq. 2.12, Eq. 2.23 gives continuous dh/dS_e near $S_e = 1$, whereas the former does not. This is generally regarded as an advantage of Eq. 2.23 over Eq. 2.12 because the continuity is consistent with observations in most cases.

Inserting Eq. 2.23 into Eq. 2.22 gives

$$k_r(S_e) = S_e^{1/2} \left[\frac{F(S_e)}{F(1)} \right]^2 \tag{2.24}$$

where

$$F(S_e) = \int_0^{S_e} \left[\frac{(x')^{1/m}}{1 - (x')^{1/m}} \right]^{1/n_{VG}} dx' \tag{2.25}$$

Substitution of $x' = (y')^{m}$ into Eq. 2.25 yields

$$F(S_e) = m \int_0^{S_e^{1/m}} (y')^{m-1+1/n_{VG}}[1-(y')]^{-1/n_{VG}} dy' \tag{2.26}$$

In a general case, a closed-form expression cannot be derived from Eq. 2.26. However, for a special case with $m = 1 - 1/n_{VG}$, integration of Eq. 2.26 leads to

$$F(S_e) = 1 - \left(1 - S_e^{1/m}\right)^m \quad (m = 1 - 1/n_{VG}) \tag{2.27}$$

Then Eq. 2.24 becomes (van Genuchten 1980):

$$k_r(S_e) = S_e^{1/2}\left[(1-(1-S_e^{1/m})\right]^{n_{VG}} \quad (m = 1 - 1/n_{VG};\ 0<m<1) \tag{2.28}$$

Both Brooks-Corey and van Genuchten relations between capillary pressure and water saturation (Eqs. 2.12 and 2.23) can be combined with either the Burdine (1952) model or the Mualem (1976) model. The particular combinations discussed herein are consistent with conventions in the literature; they are found to be more useful than other combinations in representing laboratory measurements under the local equilibrium condition.

2.3 Optimality Principles and the Euler-Lagrangian Equation

In the above section, the relevant results are valid only under the local equilibrium condition. In many cases, the local equilibrium may not hold, especially when fingering flow occurs in a porous medium. Fingering-flow patterns were investigated in a sand box by a number of researchers (e.g., Glass et al. 1988). The glass-walled box was filled with dry sand as homogenously as possible. Thus, the sand is approximately homogeneous at the continuum scale. Then water was introduced from the top of the sand box with an application rate less than the saturated hydraulic conductivity of the sand. In this case, gravitational fingering flow developed (e.g., Glass et al. 1988).

There are two major mechanisms for the fingering flow. One is the subsurface heterogeneity. Spatial variabilities of flow properties in a porous medium can result in nonuniform multiphase flow patterns with preferential flow paths. The second mechanism is nonlinear instability. As shown in Fig. 2.3, when the lighter air is displaced by water, water flow becomes unstable because of the density difference between water and air and therefore gravitational fingering flow develops. Preferential flow caused by gravitational fingering is the most frustrating process in

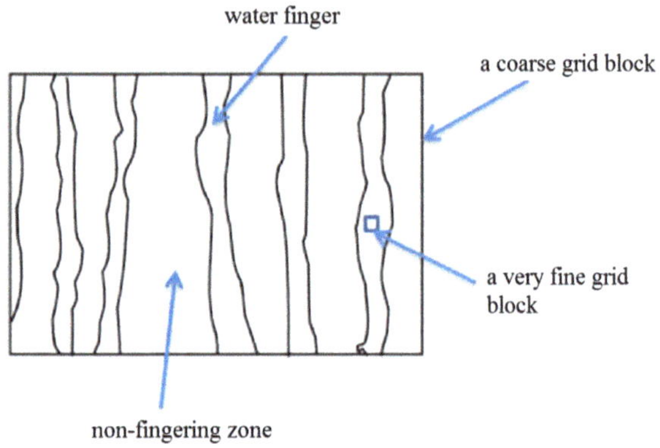

Fig. 2.3 Two scales used to model fingering flow

terms of hampering accurate prediction of contaminant transport in the vadose zone (or unsaturated soil zone) (Sheng et al. 2009). When a fluid is displaced by another less viscous fluid, viscous fingering flow will occur, which is a major technical issue in the field of oil and gas reservoir engineering, because viscous fingering significantly impacts oil recovery rate and the efficiency for oil to be displaced in a reservoir. For a real-world problem, fingering flow is really caused by a combination of the two mechanisms (Sheng et al. 2009). In this chapter, we focus on gravitational fingering in homogeneous media for a water-air flow system. The methodology and relevant results, however, can be further extended to other multiphase flow systems.

There are generally two approaches to model the fingering flow in the subsurface (Fig. 2.3). One is fine-grid simulation in which the size of a numerical gridblock is much smaller than the width of a fingering flow path. In this case, the local equilibrium condition approximately holds on the gridblock scale such that the classic Darcy-Buckingham law is valid. However, the computational effort is very intensive in this case, and it is not feasible for field-scale three-dimensional problems. Also, the required measurement problem is untenable. The other approach is to use relatively coarse grid systems in which a numerical gridblock may include one or more flow fingers. Obviously, the local equilibrium condition no longer holds because in a grid block capillary pressures are different within and outside fingering paths, while the coarse-grid simulation is more useful for practical applications. A new theory for the coarse-grid simulations that relaxes the local equilibrium condition is needed. The focus of this and following sections is on the development of such a theory for a water-air flow system (Liu 2011a). The development is based on an optimality principle that the water flow pattern in a natural unsaturated porous medium self-organizes in such a way that the total water flow resistance is minimized.

Optimality principles refer to the state of a physical process that is controlled by an optimal condition subject to physical and resource constraints (Liu 2011a). These principles have been used in many different areas, including evolution of vegetation coverage under water-limited conditions (Eagleson 2002; Liu 2011c), tree-like paths for liquid flow and heat transfer (Bejan 2000), and application of the maximum entropy production principle, in a heuristic sense, to the prediction of steady states of a wide range of systems (e.g., Tondeur and Kvaalen 1987; Bejan and Tondeur 1998; Nieven 2010; Kleidon 2009). However, the theoretical connections between these optimality principles and the currently existing fundamental laws are not fully established. Bejan (2000) argued that these principles are actually self-standing and do not follow from other known laws. In our opinion, these principles are probably a result of the characteristics of chaotic (nonlinear dynamical) systems. For a chaotic system, details of system behavior on a small scale are not predictable. However, emergent patterns, as a result of self-organization, often occur on macroscopic scales. More importantly, these patterns are self-organized in such a way that they are efficiently adapted to conditions of the relevant environment (Heylighen 2008). This adaption feature may correspond to the optimality principles. Most recently, Liu (2014) made an effort to establish a thermodynamic hypothesis for the related optimality principles, which will be discussed in Chap. 4.

The potential role of optimality principles in the formation of complex natural patterns has been recognized for many years in the surface hydrology community (Leopold and Langbein 1962; Howard 1990; Rodriguez-Iturbe et al. 1992; Rinaldo et al. 1992; Liu 2011b). For example, Leopold and Langbein (1962) proposed a maximum entropy principle for studying the formation of landscapes. Rodriguez-Iturbe et al. (1992) postulated principles of optimality in energy expenditure at both local and global scales for channel networks. However, application of these principles in the area of subsurface fluid flow has been very limited, probably because flow patterns in the subsurface are difficult to observe and characterize as a basis to motivate research activities on related emergent patterns.

In fact, optimality is really part of our daily life. As shown in Fig. 2.4, a normal person would take a straight walking path from A to B in an open space if there is not anything to be avoided between the two locations. That is the optimal path which takes the minimum energy and shortest walking time. If someone takes the relatively random path in Fig. 2.4, he or she may either be drunk or crazy because much more time and energy will be expended. However, the relatively random pathway is not irrelevant to our practice to model water flow process. If all the water elements (or "particles") travel relatively randomly through a porous medium, the overall flow pattern should be relatively uniform and fingering would not occur. For example, a diffusion process is characterized by random walks of "particles", but generates relatively uniform or smooth transport patterns.

A useful mathematical tool for dealing with optimality problems is the calculus of variations that will be used in some sections of this chapter. While readers are encouraged to consult the relevant mathematical books on the calculus of variations (e.g., Weinstock 1974), the rest of this subsection will briefly review some basic calculus results for the convenience of readers who are not familiar with the calculus of variations.

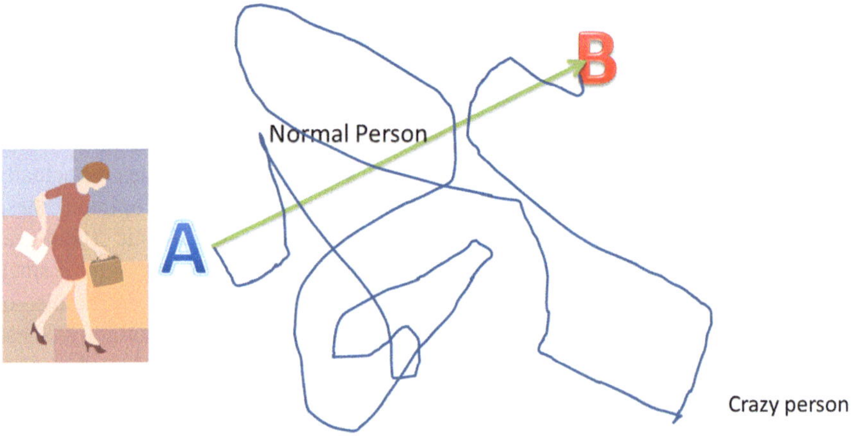

Fig. 2.4 Optimal and relatively random walk paths between points *A* and *B*

The calculus of variations studies functionals that are defined as functions of one or more functions. For example, I_y in Eq. 2.29 is a functional that is a function of unknown function y with x being an independent variable:

$$I_y = \int_{x_1}^{x_2} L(x, y, y')dx \tag{2.29}$$

where L is called a Lagrangian and is a known function for a given problem, and $y' = \mathrm{d}y/\mathrm{d}x$. The major purpose of the calculus of variations is to determine the unknown function y by solving an optimization problem for the functional I_y. (Note that x and y are used as spatial coordinates in other parts of this chapter, but here refer to an independent and a dependent variable, respectively.)

We first introduce a parameter ε and an arbitrary smooth function $\varsigma(x)$ and then construct a function like that in Fig. 2.5

Fig. 2.5 Functions $y(x)$ and $Y(x)$

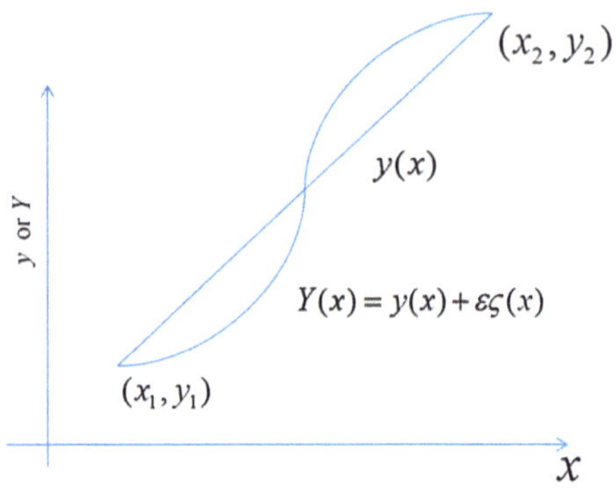

$$Y = y(x) + \varepsilon\varsigma(x) \tag{2.30}$$

where $y(x)$ is considered the solution to Eq. 2.29 when I_y is at its maximum or minimum. We further consider that $\varsigma(x)$ satisfies the following boundary condition such that Y would take the same values as y at boundaries:

$$\varsigma(x_1) = \varsigma(x_2) = 0 \tag{2.31}$$

Equation 2.30 also leads to:

$$Y' = y'(x) + \varepsilon\varsigma'(x) \tag{2.32}$$

Replacing y with Y in Eq. 2.29 and using Eq. 2.32, we have

$$\frac{dI_y}{d\varepsilon} = I_y'(\varepsilon) = \int_{x_1}^{x_2} \left(\frac{\partial L}{\partial Y}\frac{\partial Y}{\partial \varepsilon} + \frac{\partial L}{\partial Y'}\frac{\partial Y'}{\partial \varepsilon}\right)dx = \int_{x_1}^{x_2} \left(\frac{\partial L}{\partial Y}\varsigma + \frac{\partial L}{\partial Y'}\varsigma'\right)dx \tag{2.33}$$

When I_y reaches its optimum (minimum or maximum) value, Y should approach y (or $\varepsilon \to 0$) and the derivative given in Eq. 2.33 should be zero. Thus, from Eq. 2.33, we obtain

$$I_y'(0) = \int_{x_1}^{x_2} \left(\frac{\partial L}{\partial y}\varsigma + \frac{\partial L}{\partial y'}\varsigma'\right)dx = 0 \tag{2.34}$$

Integrating the 2nd term by parts gives

$$I_y'(0) = \frac{\partial L}{\partial y'}\varsigma|_{x_1}^{x_2} + \int_{x_1}^{x_2} \left[\frac{\partial L}{\partial y} - \frac{d}{dx}\left(\frac{\partial L}{\partial y'}\right)\right]\varsigma dx = 0 \tag{2.35}$$

Equation 2.31 yields the zero value for the first term on the right hand side of Eq. 2.35:

$$\frac{\partial L}{\partial y}\zeta\bigg|_{x_1}^{x_2} = 0 \tag{2.36}$$

Given the fact that ς is an arbitrary function, Eq. 2.35 leads to the well-known Euler-Lagrangian equation

$$\frac{\partial L}{\partial y} - \frac{d}{dx}\left(\frac{\partial L}{\partial y'}\right) = 0 \tag{2.37}$$

The Euler-Lagrangian equation (Eq. 2.37) is a key result for the calculus of variations and has been widely employed in optimizing a functional.

Equation 2.37 was obtained only for problems involving one independent variable. For more independent variables, a functional I_w can be defined as

$$I_w = \iiint\limits_{\Omega} L(x, y, z, w, w_x, w_y, w_z)dxdydz \tag{2.38}$$

where x, y, z are independent variables, w is an unknown function, and w_x, w_y, and w_z are derivatives of w with respect to x, y and z, respectively. Following the same procedure used to derive Eq. 2.37, the corresponding Euler-Lagrangian equation is given as (e.g., Weinstock 1974)

$$\frac{\partial L}{\partial w} - \frac{\partial}{\partial x}\left(\frac{\partial L}{\partial w_x}\right) - \frac{\partial}{\partial y}\left(\frac{\partial L}{\partial w_y}\right) - \frac{\partial}{\partial z}\left(\frac{\partial L}{\partial w_z}\right) = 0 \tag{2.39}$$

2.4 Generalization of the Darcy-Buckingham Law Based on an Optimality Condition

As previously indicated, the Darcy-Buckingham law was developed under the local equilibrium condition. This condition sometimes does not hold, especially when subgrid-scale fingering occurs. (The subgrid scale refers to the scale below a numerical grid block.) In this case, a new theory that is not restricted by the local equilibrium condition is needed. This section presents a theory of this kind based on the optimality principle that water flow pattern in an unsaturated porous medium is self-organized in such a way that the total water flow resistance is minimized (Liu 2011a).

Consider a steady-state unsaturated flow system associated with a homogeneous and isotropic porous medium. Based on the water mass conservation with a constant water density, the steady-state water flow equation is given by

$$\frac{\partial q_x}{\partial x} + \frac{\partial q_y}{\partial y} + \frac{\partial q_z}{\partial z} = 0 \tag{2.40}$$

where x and y are two horizontal coordinate axes, z is the vertical axis, and q_x, q_y and q_z are volumetric fluxes of water along x, y and z directions, respectively.

We use E (a function of x, y and z) to represent the energy of per unit weight of liquid water in soils. When kinetic energy is ignored (a common approximation for unsaturated flow because the water velocity is generally small), E is the same as the hydraulic head:

$$E = H = z + \frac{p_c}{\rho g} = z + h \tag{2.41}$$

where ρ is the water density assumed to be a constant, and h is capillary pressure head. Accordingly, the energy expenditure rate for a unit volume of porous medium, Δ_E, can be expressed as

$$\Delta_E = \frac{\partial(q_x E)}{\partial x} + \frac{\partial(q_y E)}{\partial y} + \frac{\partial(q_z E)}{\partial z} \tag{2.42}$$

The above equation simply states that for a given unit volume of porous medium, the energy expenditure rate there is equal to the energy carried by water flowing into the volume minus the energy carried by water flowing out of the volume. Obviously, a smaller energy expenditure rate corresponds to a smaller water flow resistance.

A combination of Eqs. 2.40 and 2.42 yields

$$\Delta_E = q_x \frac{\partial E}{\partial x} + q_y \frac{\partial E}{\partial y} + q_z \frac{\partial E}{\partial z} \tag{2.43}$$

Throughout this development, the *form* of the Darcy-Buckingham law is still assumed to apply to unsaturated flow:

$$q_x = -K_{un} \frac{\partial E}{\partial x} \tag{2.44-1}$$

$$q_y = -K_{un} \frac{\partial E}{\partial y} \tag{2.44-2}$$

$$q_z = -K_{un} \frac{\partial E}{\partial z} \tag{2.44-3}$$

where the unsaturated hydraulic conductivity K_{un}, unlike that in the traditional Darcy-Buckingham law (e.g., Sect. 1.2), is proposed to be (Liu 2011a)

$$K_{un} = Kk_r(h, S_*) \tag{2.44-4}$$

$$S_* \equiv \left(\frac{\partial E}{\partial x}\right)^2 + \left(\frac{\partial E}{\partial y}\right)^2 + \left(\frac{\partial E}{\partial z}\right)^2 \tag{2.44-5}$$

The relative permeability k_r is assumed to be a function of both capillary pressure head (h) and the square of the energy gradient (S_*). Assuming k_r to be a function of water flux is equivalent to assuming it to be a function of hydraulic head gradient, because water flux, hydraulic head gradient and K_{un} are related through the Darcy-Buckingham law. This assumption is used herein to take into consideration the non-equilibrium flow condition, as will be demonstrated later in this chapter. Our theory is developed for a macroscopic scale (or coarse numerical gridblock) that may include a number of subgrid fingering or preferential flow paths (Fig. 2.3). The unsaturated flow process on a local scale is mainly controlled by the pore-scale physics, while the macroscopic water-flow pattern is determined by the optimality (or the minimization of global energy expenditure rate). The local scale here refers to the continuum scale within each finger.

When we combine Eqs. 2.43 and 2.44-1 to 2.44-5, the global energy expenditure rate throughout the water-flow domain Ω is given as (Liu 2011a)

$$\iint\limits_{\Omega} \Delta_E dxdydz = \iint\limits_{\Omega} (-K_{un}S_*)dxdydz \tag{2.45}$$

The optimality principle in this study is to minimize the absolute value of the above integral.

Based on Eqs. 2.44-5 and 2.45, the Lagrangian for the given problem is

$$L = -K_{un}S_* + \lambda_L\left[S_* - \left(\frac{\partial E}{\partial x}\right)^2 - \left(\frac{\partial E}{\partial y}\right)^2 - \left(\frac{\partial E}{\partial z}\right)^2\right] \tag{2.46}$$

Note that the first term on the right hand side of the above equation is from Eq. 2.45 and the other term is a constraint from Eq. 2.44-5. The Lagrange multiplier λ_L is a function here. The constraint related to the water mass conservation, Eq. 2.40, is not included in Eq. 2.46, but, for the mathematical convenience, will be handled later. (Also for the convenience of readers who are not familiar with the mathematical background of the Lagrange multiplier, an appendix to this chapter provides an alternative derivation of Eq. 2.48 without using the Lagrange multiplier.)

The Euler Lagrangian equation (Eq. 2.39) is used to determine unknown functions associated with L to minimize the integral defined in Eq. 2.45. Equation 2.39 was derived for a single unknown function and under non-constraint conditions. However, it can be shown that Eq. 2.39 is also applied to problems with multiple unknown functions and constraints. To do so, the corresponding Lagrangian must incorporate constraints with Lagrange multipliers and the equation is applied to each unknown function separately (e.g., Weinstock 1974). Note that the use of Eq. 2.39 requires fixed-value boundary conditions which, however, can be satisfied by the fact that observed boundary conditions correspond to the optimal flow processes and are considered fixed for the steady-state flow problems under investigation here.

Replacing w with S_* in the Euler-Lagrangian equation (Eq. 2.39) yields

$$\lambda_L = \frac{\partial(K_{un}S_*)}{\partial S_*} \tag{2.47}$$

Replacing w in the Euler-Lagrangian equation with h (or E) in Eq. 2.46 and using Eq. 2.47 and the continuity equation (Eq. 2.40), we have (Liu 2011a)

$$\frac{\partial\left(\frac{\partial K_{un}}{\partial(\log S_*)}\frac{\partial E}{\partial x}\right)}{\partial x}+\frac{\partial\left(\frac{\partial K_{un}}{\partial(\log S_*)}\frac{\partial E}{\partial y}\right)}{\partial y}+\frac{\partial\left(\frac{\partial K_{un}}{\partial(\log S_*)}\frac{\partial E}{\partial z}\right)}{\partial z}=\frac{S_*}{2}\frac{\partial K_{un}}{\partial h} \tag{2.48}$$

To avoid the trivial solution of zero water flux (resulting in zero energy expenditure), we need to consider an additional constraint that the magnitude of water flux averaged over the whole flow domain, $\overline{q}$, is non zero and fixed. In general, it is difficult to obtain an analytical solution to Eq. 2.48. To obtain a practically useful closed-form result, we introduce the following approximation: $\frac{S_*}{2}\frac{\partial K_{un}}{\partial h}=\frac{S_*^{1/2}}{2}\frac{\partial(K_{un}S_*^{1/2})}{\partial h}=\frac{S_*^{1/2}}{2}\frac{\partial q}{\partial h}\approx\frac{S_*^{1/2}}{2}\frac{\partial\overline{q}}{\partial h}=0$, where q is the magnitude of water flux and given by Eq. 2.53. Note that in the above approximation, we use the magnitude of average water flux to represent the local value on the right-hand side term of Eq. 2.48. This approximation should be reasonable for relatively uniform water-flux fields on the macroscopic scale. In more general cases, the effect of this approximation is expected to be absorbed by the fitting parameter a_0 in the final solution (Eqs. 2.54-1 and 2.54-2).

A comparison between Eq. 2.48 (without the term on the right hand side) with the continuity equation (Eqs. 2.40 and 2.44-1 to 2.44-5) yields

$$\frac{\partial K_{un}}{\partial(\log S_*)}=A'K_{un} \tag{2.49}$$

where A' is a constant.

Furthermore Liu (2011a) considers K_{un} to be approximately expressed by

$$K_{un}=Kk_r(h,S_*)=Kf_h(h)g_s(S_*) \tag{2.50}$$

where f_h and g_s are functions of h and S_*, respectively. Substituting Eq. 2.50 into Eq. 2.49 results in (for a given location)

$$g_s(S_*)\propto S_*^{A'} \tag{2.51}$$

Based on the Darcy-Buckingham law, Eq. 2.51 can be rewritten as

$$g_s(S_*)\propto\left(\frac{q}{K_{un}}\right)^{A'/2} \tag{2.52}$$

where q is the magnitude of water flux given by

$$q=\left[q_x^2+q_y^2+q_z^2\right]^{1/2} \tag{2.53}$$

Combining Eqs. 2.50 and 2.52 gives the final conductivity relationship of Liu (2011a) as follows

$$K_{un} = KF_h(h)\left(\frac{q}{K}\right)^{a_0} \quad (2.54\text{-}1)$$

where a_0 is a constant. Equation 2.54-1 may be rewritten as

$$k_r(h, S_*) = F_h(h)\left(\frac{q}{K}\right)^{a_0} \quad (2.54\text{-}2)$$

There may be different interpretations of Eqs. 2.54-1 and 2.54-2. One interpretation is that $F_h(h)$ is the local-scale relative permeability within the fingering-flow zone, such as those discussed in Sect. 2.2, and that the power function of flux in the equation represents the fraction of fingering flow zone in the cross section normal to the water flux direction. (Note that the local scale here refers to the continuum scale within each finger, as previously indicated.)

The validity of Eqs. 2.54-1 and 2.54-2 for homogeneous soils was demonstrated with laboratory-experimental observations of vertical fingering flow (Wang et al. 1998). After compiling and analyzing a number of data sets for fingering flow in sand boxes, Wang et al. (1998) empirically found that the fraction of fingering flow zone (or effective water-flow area), f, is given as

$$f = \left(\frac{q}{K}\right)^{0.5} \quad (2.55)$$

Obviously, this empirical relation is consistent with our theoretical result (Eqs. 2.54-1 and 2.54-2) with $a_0 = 0.5$. At this point, our theory is not able to give a definite value for parameter a_0; whether a universal a_0 value exists for water flow in unsaturated porous media is an issue needing further research.

Equations 2.54-1 and 2.54-2 clearly shows that for a gravity-dominated unsaturated flow process, relative permeability is not only a function of water capillary pressure head (or saturation), but also a power function of water flux. Treating relative permeability as a function of fluid capillary pressure head (or saturation) only has been used widely in the fields of vadose zone hydrology, soil sciences, and reservoir engineering. This needs to be revisited, because the treatment is based on the local-equilibrium assumption that capillary pressure is uniform within a numerical gridblock. This assumption is obviously violated at a large scale by the existence of fingering flow (Fig. 2.3), as previously indicated. The power-function term in Eqs. 2.54-1 and 2.54-2 largely reflects the self-organization of flow patterns driven by the minimization of total flow resistance, or the global energy expenditure rate. In fact, Eqs. 2.54-1 and 2.54-2 also provides a method to relate local-scale laboratory measurement [or $F_h(h)$] to large-scale hydraulic properties (Eq. 2.54-2).

An interesting analogue of our new development is a relationship for viscous fluid flow in the fluid mechanics literature. In Table 2.1, in the upper left hand

Table 2.1 A scientific analogue

$\tau_w = \mu \frac{\partial v_w}{\partial y}$	$q = -K_{un} \frac{\partial H}{\partial x}$
Laminar flow: viscosity μ is a water property and independent of velocity	**"Uniform" flow**: K_{un} is a function of saturation and independent of water flux
Turbulent flow: viscosity depends on velocity (through *Re* number)	**Fingering flow**: K_{un} is a function of water flux

corner is a well-known relationship between shearing stress (τ_w) and velocity gradient for water (a Newtonian fluid), where v_w is velocity parallel to the stress direction, μ again is viscosity, and y here is a spatial axis normal to the stress. When water flow is laminar (stable), the viscosity, corresponding to K_{un} in this work, is a water property and independent of the velocity. However, when water flow becomes turbulent (unstable), the apparent viscosity is related to Reynolds number (*Re*) that is a function of water velocity. Note that the turbulent-flow case corresponds to our new expression for the conductivity because both cases are associated with unstable flow processes. This analogue highlights the needs to develop different theories for different flow regimes. While the classic Darcy-Buckingham law was developed based on the local equilibrium assumption, the new theory intends to deal with water flow when the assumption is not valid anymore.

While there are a number of studies on applications of optimality in the literature, the study of Liu (2011a) is unique in revealing that the unsaturated hydraulic conductivity is a power function of water flux. This interesting finding has several important implications. Firstly, it makes sense within the context of resource (or conductance) allocation. The power-function relationship with a positive exponent value always gives a relatively small flow resistance at a location with a large flux, such that flow in the whole system is the most efficient. This allocation strategy is also consistent with our daily-life experience. For example, in a highway system, locations with high traffic flux are generally wider or have larger conductivities. Secondly, while complex partial differential equations are involved in the derivation procedure discussed above, the form of the final results (a power function) is very simple. This form likely has something to do with fractal patterns, involving nonlinear dynamics, that have been observed and studied intensively in the literature (e.g., Feder 1988). In general, a fractal has many features that can be characterized by power functions and is related to chaotic systems. A detailed exploration of a possible linkage between our finding and fractals is beyond the scope of this book and left to future investigations. Thirdly, one grand challenge facing us in the area of subsurface flow is the need to develop physical laws for large-scale multiphase-flow problems. At a local scale smaller than the finger width, fluid distribution is mainly controlled by capillarity and not sensitive to flow conditions. That is why relative permeability at a local scale can be successfully described as a function of saturation (or capillary pressure) only. At a large scale, this is not the case anymore, although local-scale relationships have been widely used at large scales because alternatives are unavailable. It is fair to say that as a result of the high nonlinearity involved, how to model large-scale multiphase flow is an issue that has not been resolved at a fundamental level. The result here (Eqs. 2.54-1 and 2.54-2) suggests that function forms of large-scale relationships to describe multiphase flow are very likely different from their counterparts at the local scale, which cannot be resolved by upscaling parameters based on the same function forms as those at local scales. It is our hope that the optimality approach may provide an important way to obtain such large-scale relationships.

We need to indicate that Eqs. 2.54-1 and 2.54-2 can be rewritten in some approximate forms that are more convenient for practical applications. For example, for gravitational fingering flow, gravity is the dominant driving force for vertical flow and capillary pressure gradient can be practically ignored. In this case, hydraulic gradient equals to negative one (Eqs. 2.44-1 to 2.44-5), and Eqs. 2.44-1 to 2.44-5 becomes

$$q = K_{un} = K_a f \tag{2.56}$$

where K_a is the average (unsaturated) hydraulic conductivity within the fingering flow region (or the active flow region that is denoted by the subscript a), and f again is volumetric fraction of a porous medium occupied by fingering flow and is given by

$$f = \left(\frac{q}{K}\right)^{a_0} \tag{2.57}$$

The above equation is a direct result of Eq. 2.54-2. Also note that in Eq. 2.56, the multiplication of K_a by f takes into consideration that only a fraction of a porous medium actually involves water flow while the rest is simply bypassed.

We further apply the Brooks-Corey (1964) relation (Eq. 2.13) to water flow within the fingering flow region:

$$\frac{K_a}{K} = S_a^{\beta_0} = (S_e/f)^{\beta_0} \tag{2.58}$$

where S_a is effective water saturation within the fingering flow region, β_0 is a constant, and S_e is defined in Eq. 2.7. Combining Eqs. 2.56–2.58 yields

$$f = S_e^{\gamma} \tag{2.59}$$

where

$$\gamma = \frac{a_0\beta_0}{1 + a_0\beta_0 - a_0} \tag{2.60}$$

To briefly demonstrate the practical importance of the generalized Darcy-Buckingham law, we use an ideal example to show differences between calculation results from the classic Darcy-Buckingham law and its generalized version. Consider a one-dimensional vertical water flow process in initially dry and homogeneous soil (with zero residual water content); the flow is gravity dominated. Then we estimate the water pore velocity (v) under the steady-state flow condition for a fixed water flux imposed on the top surface of the soil:

$$v = \frac{q}{\phi S_e} \tag{2.61}$$

where ϕ is soil porosity. By definition, we can obtain the dimensionless water flux q' as

$$q' = \frac{q}{K} = \frac{K_a}{K} f \tag{2.62}$$

Combining Eqs. 2.58, 2.59 and 2.62 gives:

$$S_e = (q')^{\frac{1}{\beta_0 + \gamma(1-\beta_0)}} \tag{2.63}$$

Then the normalized pore velocity $\overline{v}$, defined as the pore velocity divided by the pore velocity without fingering flow (or $\gamma = 0$), is:

$$\overline{v} = (q')^{\left(\frac{1}{\beta_0} - \frac{1}{\beta_0 + \gamma(1-\beta_0)}\right)} \tag{2.64}$$

Figure 2.6 shows the normalized water pore velocity as a function of γ, calculated with Eq. 2.64, for $\beta_0 = 4$ and for $q' = q/K = 0.1$ and 0.3, respectively. There are several interesting observations that can be made from Fig. 2.6. Firstly, as expected, the pore velocity calculated with the generalized Darcy-Buckingham law ($\gamma > 0$) can be significantly larger than that calculated with the classic

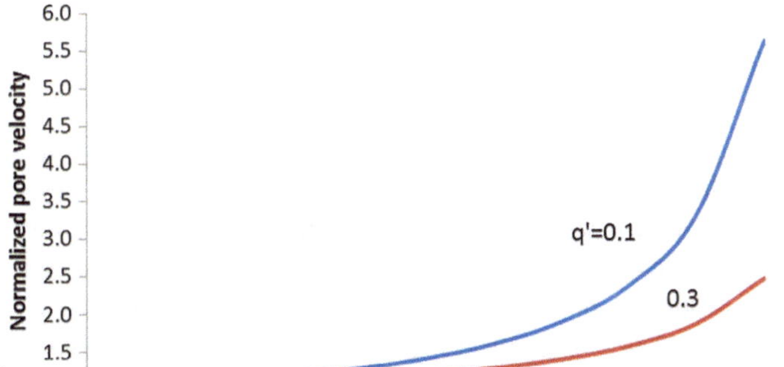

Fig. 2.6 Normalized water pore velocity as a function of γ, calculated with Eq. 2.64, for $\beta_0 = 4$ and for $q' = q/K = 0.1$ and 0.3, respectively. The pore velocity calculated with the Darcy-Buckingham law corresponds to a normalized pore velocity equal to one

Darcy-Buckingham law. This is because the latter is based on the local-equilibrium assumption and could not theoretically handle fingering-flow effects. Underestimation of water pore velocity (or contaminant advection velocity) has a significant consequence for environmental assessment of groundwater contamination originating from the ground surface (e.g., Sheng et al. 2009). Secondly, for a given water flux, the pore velocity is a strong function of parameter γ. A larger γ generally corresponds to a smaller water saturation S_e (Eq. 2.63), and consequently to a larger pore velocity. Note that in an extreme case, $\gamma = 1$ gives water saturation within the fingering flow region $S_a = S_e/f = S_e^{1-\gamma} = 1$. In this case, the fingering flow region is fully water-saturated and thus the corresponding pore velocity is the highest for a given water flux. Thirdly, for a given γ value, the normalized pore velocity decreases with increasing water flux. This is because a larger water flux enhances the fingering flow region and therefore generally reduces the non-uniformity of the water flow pattern.

2.5 Verification with Field Observations of Unsaturated Water Flow in Soils

In Sect. 2.4, we have presented a new theory for liquid water flow in unsaturated media, or the generalized Darcy-Buckingham law, based on an optimality principle. Its consistence with laboratory experiments reported by Wang et al. (1998) is discussed there as well. The focus of this section is on a further verification of the theory using field-scale data related to gravitational fingering flow in unsaturated soils. The new theory is now called the active region model (ARM) in the literature of modeling unsaturated water flow in soils (Liu et al. 2005). The contents of this section are mainly based on the work by Sheng et al. (2009).

2.5.1 Field Experiments

For verifying the ARM, dyed-water infiltration tests were conducted in unsaturated soils at two different field sites under various hydraulic conditions, from June 2006 to Oct. 2007. For each test, dyed water was applied from the soil surface, which allows for determination of water flow patterns by visualizing dye patterns. Site I (Fig. 2.7) is located in the Irrigation and Drainage Station of Wuhan University, China, and the corresponding soil is a quite homogeneous loam. Site II (Fig. 2.7) is located at the foot of Mountain Luojia, China. Because macropores were observed in the top soil layer at Site II, the top 40 cm soil layer was removed before performing the tests, such that the tests did not involve water flow in macropores to simplify the flow process under investigation. Soil below the depth of 40 cm at Site II consists of unstructured clay without macropores. Soil samples were collected at each site from the depth intervals of 0–10, 10–20, 20–50 and 50–100 cm.

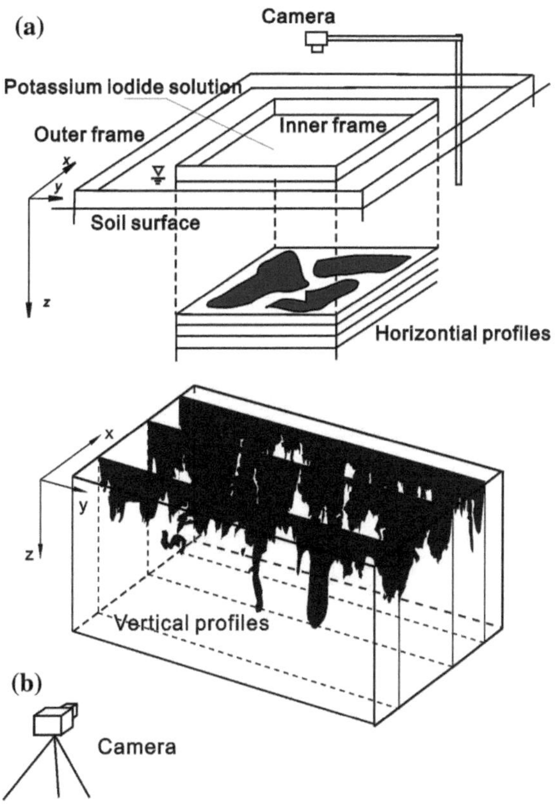

Fig. 2.7 Experimental set-ups: **a** site I and **b** site II (Sheng et al. 2009) (reproduced by permission of Elsevier)

Then soil properties were measured for these samples, including the soil texture, bulk density, porosity, and saturated hydraulic conductivity (Table 2.2). The soil water retention function for each site was estimated from soil texture; the residual water contents are 0.07 and 0.12 for the loam (Site I) and clay (Site II) soils, respectively.

Seven distinct dyed-water infiltration tests were conducted at Site I (plots 1 through 7) and six tests at Site II (plots 8 through 13) (Table 2.3). A test location at a given site is called a test plot here. The experimental design is shown schematically in Fig. 2.7 for each plot, including two rectangular frames embedded concentrically into the given plot. The outer frame is larger than the inner frame by 1 m in both length and width, such that lateral boundary effects can be essentially eliminated for water flow within the inner frame, below which experimental observations were made. Each plot was leveled to ensure a uniform surface (boundary) condition for the infiltration tests. Test plots at the same site were placed at least 2 m apart to avoid interference among them. Top soil surface of a plot was initially covered with a plastic film. Then, the inner frame top surface was prepared by ponding dyed water on the top of the plastic film, while the outer border area was ponded to the same depth with fresh water. When the experiment started, the

Table 2.2 Soil physical and hydraulic properties at the two test sites (Sheng et al. 2009)

Site	Depth (cm)	Texture			Bulk density (g cm^{-3})	Porosity (%)	Saturated hydraulic conductivity (cm s^{-1})
		>50 μm %	2–50 μm %	<2 μm %			
I	0–10	29.4	49.2	21.4	1.38	38.8	3.4×10^{-4}
	10–20	28.2	49.4	22.4	1.40	40.2	1.2×10^{-4}
	20–50	31.3	46.2	22.5	1.44	40.6	1.1×10^{-4}
	50–100	32.1	44.8	23.1	1.45	44.4	1.1×10^{-4}
II	0–10	3.5	55.0	44.5	1.44	40.2	5.8×10^{-5}
	10–20	4.4	51.5	44.4	1.50	42.7	2.1×10^{-5}
	20–50	4.3	51.4	44.3	1.50	44.8	2.2×10^{-5}
	50–100	4.7	50.7	44.6	1.59	45.1	1.4×10^{-5}

Table 2.3 The dye infiltration conditions of each plot (Sheng et al. 2009)

Ponding water depth (mm)	Loam		Clay		
	100 × 100 cm^2		100 × 100 cm^2	100 × 200 cm^2	100 × 400 cm^2
20	Plot 1	Plot 2	Plot 8	/	/
40	Plot 3	Plot 4	Plot 9	Plot 12	Plot 13
60	Plot 5	Plot 6	Plot 10	/	/
80	Plot 7	/	Plot 11	/	/

plastic film was removed immediately, thus creating an almost instantaneous ponding infiltration from both inner and border frames. Then the experimental plot was covered to prevent evaporation and left for 12 h to complete the infiltration process (Yasuda et al. 2001; Wang et al. 2006).

Twelve hours after the dyed water was applied at the top boundary, the flow pattern visualization was started for soil surfaces created during excavation of the corresponding soil column. For a plot at Site I, the soil was removed horizontally layer by layer (Fig. 2.7a) with a vertical interval of 1–5 cm from top to bottom. At Site II, a trench (that is 1.0 m apart from a test plot) was manually dug to a depth of 1.5 m, one day before the dyed water was applied. After infiltration, vertical soil profiles were excavated across the plot at a horizontal interval of 5 cm. After each soil layer or vertical profile was exposed, the soil surface was leveled and cleaned with a brush to remove soil particles generated from digging. The dyed (or "stained") flow patterns in the horizontal (Site I) and vertical (Site II) soil slices were recorded during day time using a CCD digital camera. Then soil samples were collected from both the stained regions and the unstained regions in a soil slice to measure the soil water content distributions.

2.5.2 Data Analysis Methods

The new theory is given in Eqs. 2.54-1 and 2.54-2 or Eq. 2.59. The two equations are equivalent for gravity dominated flow. Herein we focus on Eq. 2.59 because it can be easily compared with field observations. In the traditional continuum approaches, the whole flow region is considered to be conductive, the flow and transport occur in the whole flow region, and all the parameters used in the governing equations are related to the whole flow region. The main idea behind the new theory is that the flow domain can be divided into active and inactive regions, and (fingering) flow occurs only in the active region, so that the inactive region is simply bypassed. Thus, the theory is called the active region model (ARM) for water flow in unsaturated soils (Liu et al. 2005), as previously indicated. The relative portion of the active region is dynamic and expressed as a power function of soil effective saturation, as given in Eq. 2.59 with γ being called the ARM parameter in this study. Also, Eq. 2.59 is derived under a condition that water content is at its residual value in the inactive region. When the above condition is not met and yet water is relatively immobile in the inactive region, effective saturation S_e in Eq. 2.59 should be replaced by S_e^*, the average active water saturation for the whole flow region, defined as:

$$S_e^* = \frac{V_{aw}}{V_p} \tag{2.65}$$

where V_{aw} is the total water volume (excluding the volume corresponding to the residual water content) in the active region, V_p is the pore volume (excluding the volume corresponding to the residual water content) of the whole flow region (including both the active and inactive parts). From the definition of f, S_e^*, as previously indicated, can be calculated by

$$S_e^* = fS_a \tag{2.66}$$

where the effective water saturation (within the active region) S_a is given as

$$S_a = \frac{V_{aw}}{V_{ap}} \tag{2.67}$$

where V_{ap} is the pore volume (excluding the residual pore volume) of the active region. Thus, S_a is calculated by the following equation

$$S_a = \frac{\theta_a - \theta_r}{\theta_{sat} - \theta_r} \tag{2.68}$$

where θ_a is the water content in the active region. As defined in Eq. 2.7, θ_{sat} is saturated water content and θ_r is residual water content.

We evaluate the ARM by examining (a) whether Eq. 2.59 satisfactorily represents the relevant field observations and (b) whether the parameter γ is constant with different test conditions for a given test site (Sheng et al. 2009). We also investigate how the parameter γ changes with test scale. In the ARM, the parameter γ is a positive constant (between zero and one), depending on soil properties, and is used for characterizing the fingering water flow properties (Liu et al. 2005).

2.5.3 Results and Discussion

Fingering flow patterns are generally complex, as demonstrated by dyed flow paths in Figs. 2.8, 2.9, 2.10 and 2.11. The stained or dyed region is considered to be the active region, or the location where water flow occurs. Figure 2.8 shows vertical flow patterns for selected soil slices at Plots 8, 9, 10 and 11 at Site II. Relatively homogeneous flow patterns are observed only in the upper soil layer with a thickness of 5–20 cm. The flow becomes irregular and preferential below that soil layer and bypasses a significant soil portion. The observed flow pattern also depends on test conditions, such as ponding water depth applied to the top surface. A larger water depth results in a larger average water-penetration depth. Because the water flow is three dimensional in nature, some disconnections of flow patterns are observed in Fig. 2.8 and other relevant figures that present two dimensional patterns only. Figure 2.9 shows stained patterns in horizontal planes at various depths for Plot 3 and Plot 4 at Site I. Figures 2.10 and 2.11 present vertical flow patterns for two plots (Plots 12 and 13 at Site II) with relatively large horizontal sizes (Table 2.3). While the soils at the two test sites are relatively homogeneous, the flow patterns are so irregular that it will be very difficult, if not impossible, to model all the details in practice. Within a macroscopic (or coarse-grid) modeling framework, the flow is treated as a one-dimensional process along the vertical direction. Observed flow patterns clearly show that equilibrium in a horizontal plane does not exist. Thus, the classic Darcy-Buckingham law does not apply for a coarse-grid model. This highlights the needs to relax the local-equilibrium assumption, as previously discussed.

To verify the generalized Darcy-Buckingham law, or ARM in this case, we need to demonstrate that Eq. 2.59 is consistent with the field observations. In a horizontal plane, the ratio of dyed soil area to the total soil area is called coverage of the stained region and taken to be the fingering flow zone fraction (or f) at the corresponding depth. Figure 2.12 shows the coverage of the active (stained) region (the stained region) versus depth for the plots. As previously indicated, Sheng et al. (2009) used effective saturation calculated from the water saturation in the active region, defined in Eq. 2.66, to evaluate the relationship between f and water saturation (Eq. 2.59). In the active region, all the water, excluding residual water, is assumed to be mobile. Therefore, experiments should ideally be carried out under conditions with initial soil saturation near the residual value. However, under field conditions, it is virtually impossible to carry out experiments under such conditions.

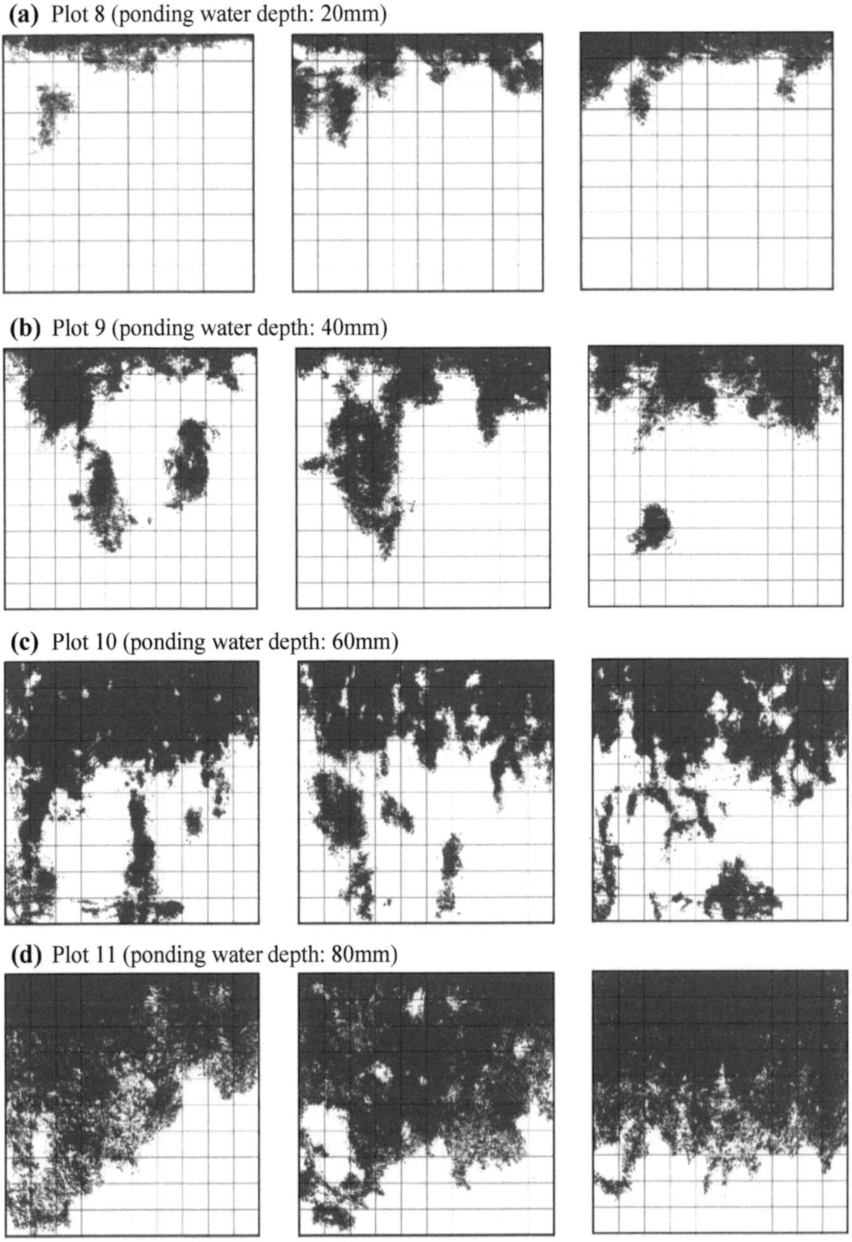

Fig. 2.8 Flow patterns for plots 8, 9, 10 and 11 at site II (Sheng et al. 2009) (reproduced by permission of Elsevier)

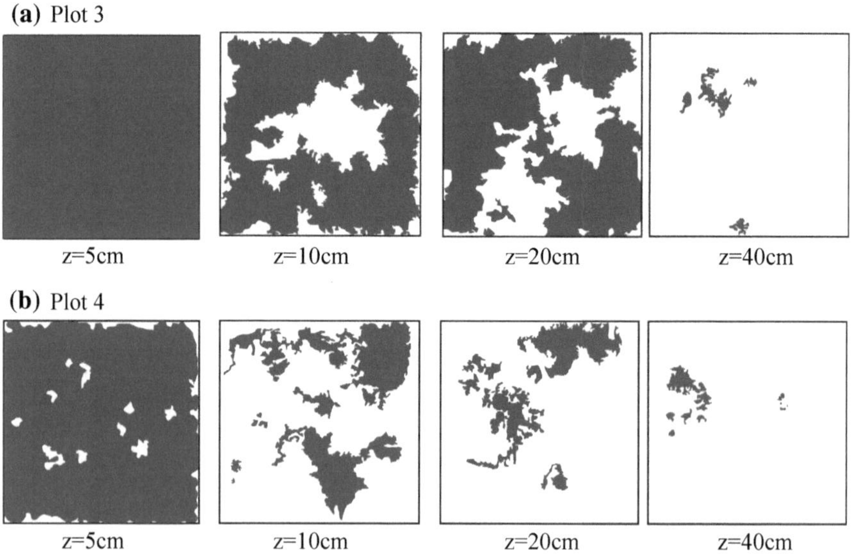

Fig. 2.9 Stained patterns at various depths (or elevations) for plot 3 and plot 4 at site I. Note that $z = 0$ on the ground surface

To analyze the test data, as previously discussed, Sheng et al. (2009) assumed that before the test, all water initially in the soil is immobile during the time periods of the tests, although the water content is higher than the residual water content. This can be justified, given the fact that water flows much more quickly in the active (stained) region. Therefore, soil water in the "inactive" region with initial water content higher than the residual content can be approximately considered immobile at least during test periods.

The relation between the effective saturation and the stained coverage for each plot is presented in Fig. 2.13. In general, Eq. 2.59 fits the test results reasonably well for a full range of water saturations under different test conditions, indicating that the complex preferential flow feature can be captured with the ARM. The fitted γ values are close for different initial water contents and different total infiltrating water depths for a given test site (Fig. 2.14). This is consistent with the ARM in which the parameter γ is constant for a given site, although the flow pattern details may change considerably with different test conditions (Figs. 2.8, 2.9, 2.10 and 2.11). The standard deviation of dyed flow depth σ is also given in the Fig. 2.14 as a measure of complexity or heterogeneity of the water flow pattern. For a given location on the soil surface, the so-called dyed flow depth is defined as the largest depth of dyed soil.

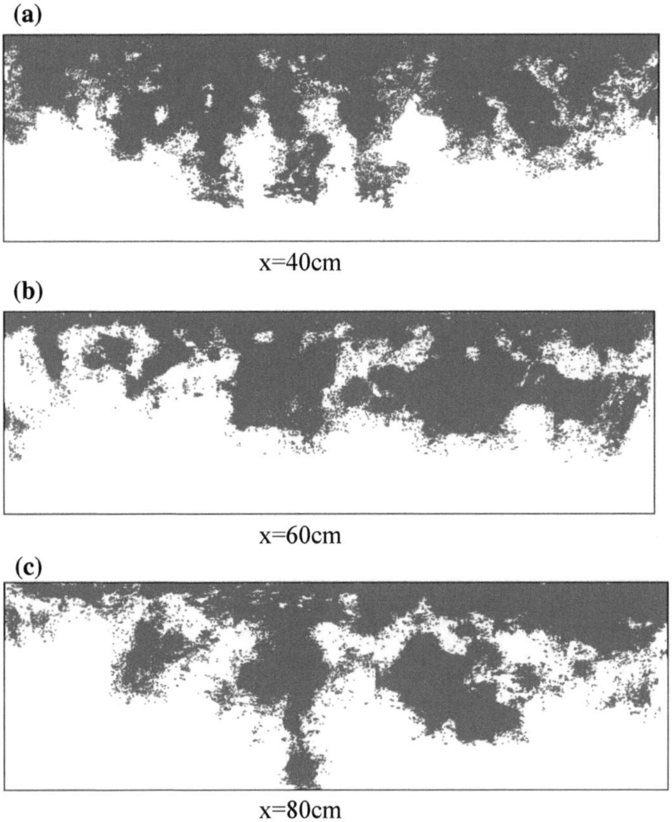

Fig. 2.10 Flow patterns in vertical cross sections for plot 12 at site II

As indicated in Fig. 2.14, while the degree of flow-pattern heterogeneity (characterized by σ) can vary considerably with test conditions (such as the ponding water depth), variations in γ values estimated from the observations are small. The flow patterns for the plots with ponding-water depths of 40 and 60 mm are most heterogeneous (characterized by the largest σ values in Fig. 2.14) for plots with a horizontal dimension of 100×100 cm^2 at both Sites I and II. To test impacts of test scales, the infiltration tests were also conducted for two relatively large plots (Plots 12 and 13 with horizontal dimensions of 100×200 cm^2 and 100×400 cm^2, respectively) (Table 2.3). The observed flow patterns for selected vertical soil slices are given in Figs. 2.10 and 2.11. The estimated γ values for these two large plots are similar to those for the smaller plots at the same site, indicating that the ARM is robust in dealing with fingering flow at different scales, although the test scale range is not large enough to obtain more conclusive results regarding this issue.

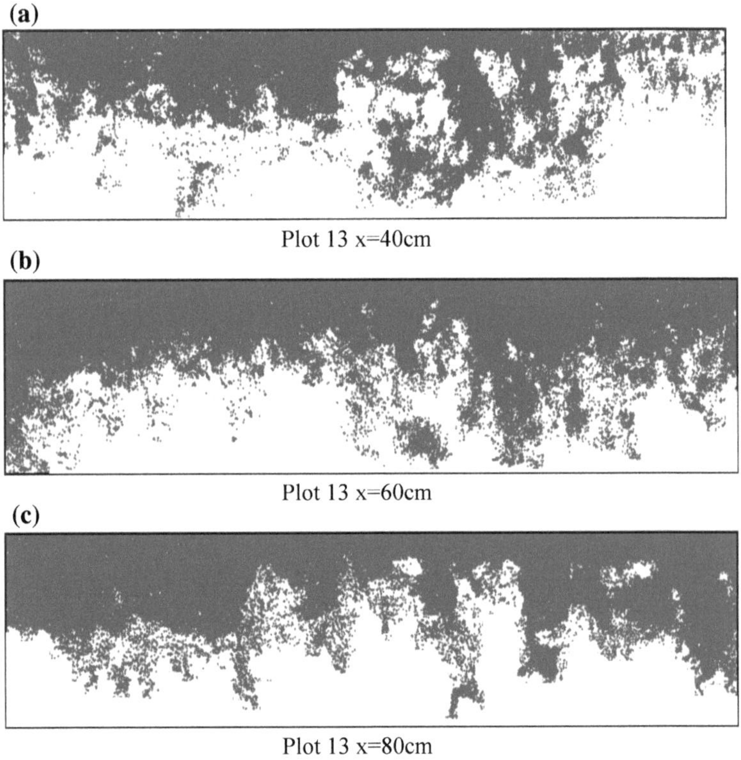

Fig. 2.11 Flow patterns in vertical cross sections for plot 13 at site II (Sheng et al. 2009) (reproduced by permission of Elsevier)

In summary, the test results provide a valuable data set for evaluating the ARM that is based on the generalized Darcy-Buckingham law (Sect. 2.4). It was found that the horizontally averaged water flow patterns under different test conditions are satisfactorily described by the ARM relation (Eq. 2.59). While details of flow patterns (or flow heterogeneity) are obviously dependent on specific test conditions, our analysis shows that the fitted values for the ARM parameter γ are not sensitive to the test conditions including the initial water content and the top boundary condition for a given test site. This further supports the validity of the ARM in which the parameter γ is constant for a given site and soil type.

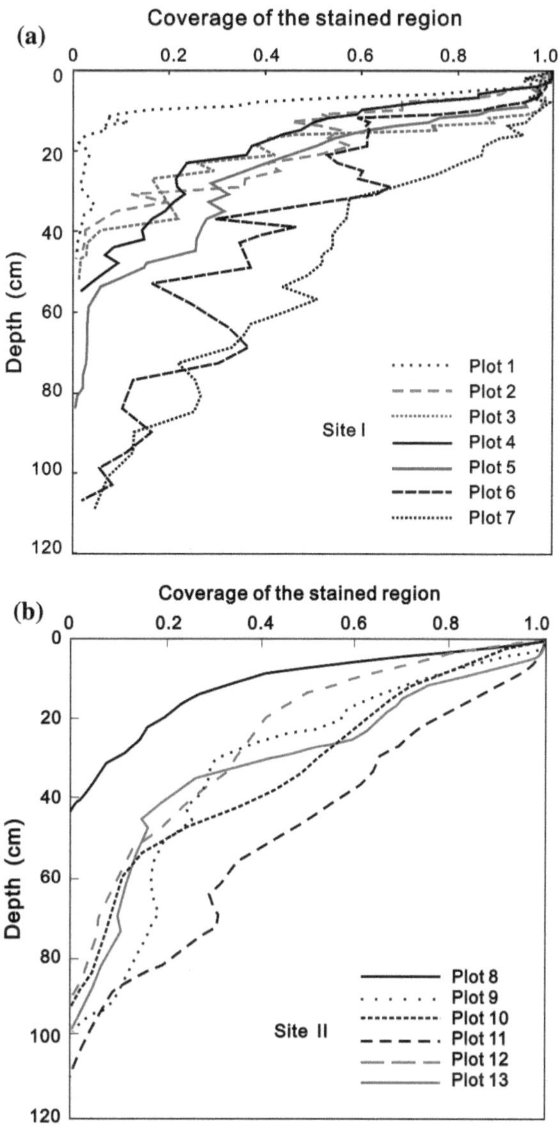

Fig. 2.12 Coverage of the stained region as a function of depth (Sheng et al. 2009) (reproduced by permission of Elsevier)

2.6 The Active Fracture Model: An Equation for a Mountain

As indicated in the preface, the work documented in this book was motivated by the needs in practical applications. Our effort of generalizing the Darcy-Buckingham law is a direct result of addressing an important technical issue for the Yucca Mountain Project: the accurate modeling of long-term flow and transport processes.

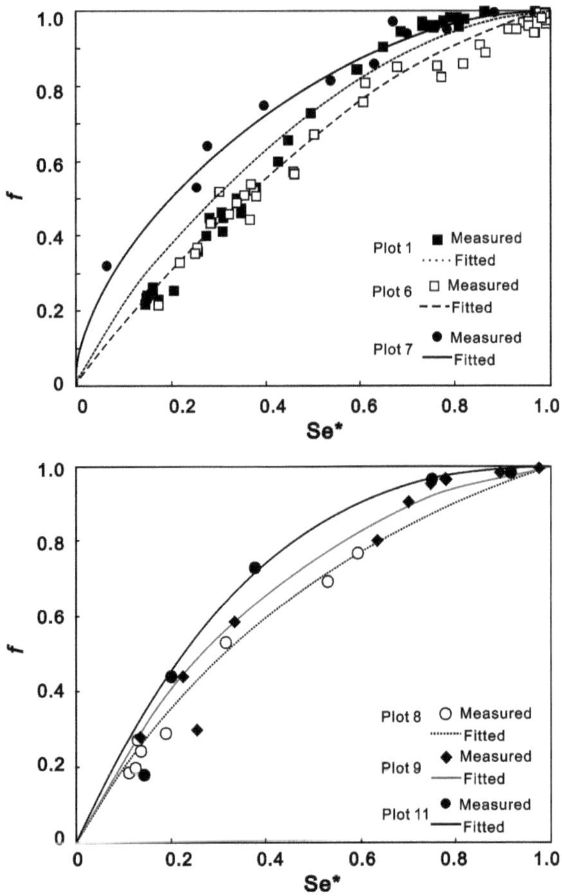

Fig. 2.13 Relationships between the soil water saturation and the coverage of stained region that are used to fit ARM parameter γ (Sheng et al. 2009) (reproduced by permission of Elsevier)

Yucca Mountain is located in Nevada, USA, and is about 90 miles from Las Vegas. It is the US's national site for geological disposal of high-level nuclear wastes. The project, given the total number of capable scientists and engineers involved, can be regarded as the "Manhattan" project in the area of subsurface science and engineering. This section briefly introduces the Yucca Mountain Project and the active fracture model that was developed for modeling flow and transport processes in Yucca Mountain and is an application of the generalized Darcy-Buckingham law.

2.6.1 Yucca Mountain Project

High-level radioactive wastes are the highly radioactive material produced as a byproduct of the reactions occurring inside nuclear reactors that generate electricity in nuclear power plants. Because of their highly radioactive fission products,

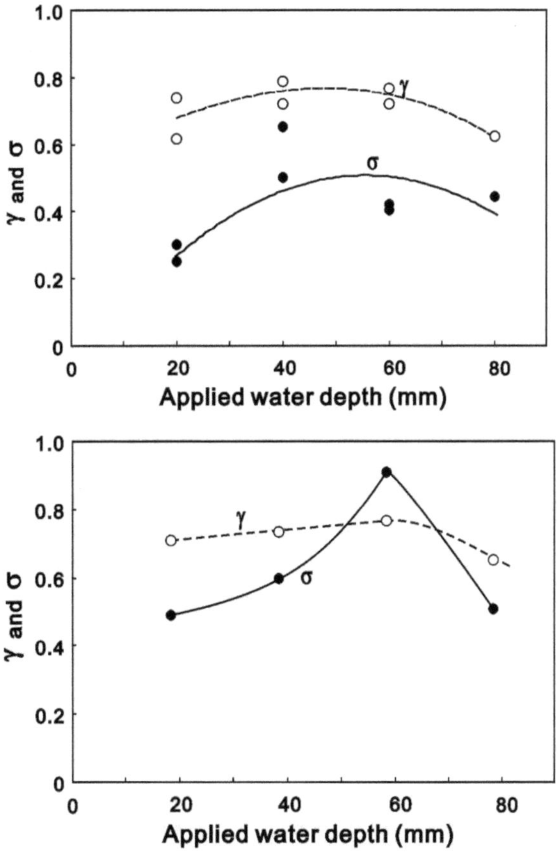

Fig. 2.14 ARM parameter γ and parameter σ as functions of infiltration water depth (Sheng et al. 2009) (reproduced by permission of Elsevier). Both parameters are related to flow-pattern heterogeneity

high-level waste and spent fuel must be handled and stored with care. There are approximately 72,500 metric tons of nuclear wastes nowadays in USA (Garrick and Di Bella 2014). Since the only way for radioactive waste to finally become harmless is through decay, which for high-level wastes can take hundreds of thousands of years, the waste must be stored and finally disposed of in a way that provides adequate protection of the public for a very long time. Geological disposal has been considered the most feasible way to manage nuclear wastes.

In 1982, the US Congress passed the Nuclear Waste Policy that is a federal law to deal with the geological disposal of high-level nuclear waste in USA. According to the law, the US Department of Energy (DOE) has the responsibilities to site, construct, operate and close a geologic repository (US DOE 2002). DOE began studying Yucca Mountain, Nevada, and other potential sites in 1978 to determine whether they would be suitable for being the first national geologic repository. In 1984, President Ronald Reagan approved three sites for further scientific investigation that is called site characterization. In 1987, the US Congress passed an amendment to the Nuclear Waste Policy Act that directed DOE to investigate the

Yucca Mountain site exclusively (US DOE 2002). The site was recommended by President George W. Bush in 2002, which ended the site characterization phase and allowed the DOE to move forward to establish the first national repository to store nuclear waste at the Yucca Mountain site. However, the Yucca Mountain Project has not been popular in Nevada. Following the 2006 mid-term congressional elections, Harry Reid, the democratic Nevada senator and a longtime opponent of the repository, became the US senate speaker. In this capacity, he has successfully and significantly cut DOE's funding for the Yucca Mountain Project. The US DOE, however, was able to complete the License Application documenting the scientific investigations and engineering designs for the site and submitted it in 2008 to Nuclear Regulatory Commission (NRC) that, in addition to its other responsibilities, serves as the regulator for the Yucca Mountain Project. During his 2008 presidential campaign, Senator Barack Obama promised to Nevada voters to cancel the Yucca Mountain project. President Barack Obama, after elected, then dismantled DOE's Office of Civilian Radioactive Waste Management that managed Yucca Mountain Project and zeroed out federal funding for the project. This caused several lawsuits against Obama administration's decision. Forced by a court ruling, NRC used the funding left over from the previous years to finish several key reports by its scientific staff, called safety analysis reports, in 2014 and 2015 to independently evaluate the scientific work documented in DOE's License Application. Conclusions of these NRC's reports generally indicate that the DOE's scientific work is valid and adequate for supporting the development of a national geological disposal repository at the Yucca Mountain site.

2.6.2 The Active Fracture Model (AFM)

Numerical models are the only tools to help determine if a geological repository can safely store high-level nuclear wastes in a way that provides adequate protection of the public for a very long time (a million year in the case of Yucca Mountain Project). Thus, the accurate modeling of long-term flow and transport processes impacting the migration of radionuclides from nuclear waste packages is a key technical issue for assessing the performance of a repository. The active fracture model (AFM) was developed by Liu et al. (1998) to address that issue for the Yucca Mountain Project and has been used as the base case theory for modeling flow and transport in the unsaturated zone of the site. This subsection briefly discusses the geological model for the unsaturated zone of Yucca Mountain and then presents the AFM, including its relation with our optimality results (Sect. 2.4) and verification with field data.

Geological formations of the unsaturated zone of Yucca Mountain were grouped into stratigraphic units, based on their degrees of welding, by Montazer and Wilson (1984). The stratigraphic units consist of the following, in descending order from the ground surface: welded Tiva Canyon Tuff (TCw); mainly nonwelded Paintbrush Group (PTn); welded Topopah Spring Tuff (TSw); mostly nonwelded and

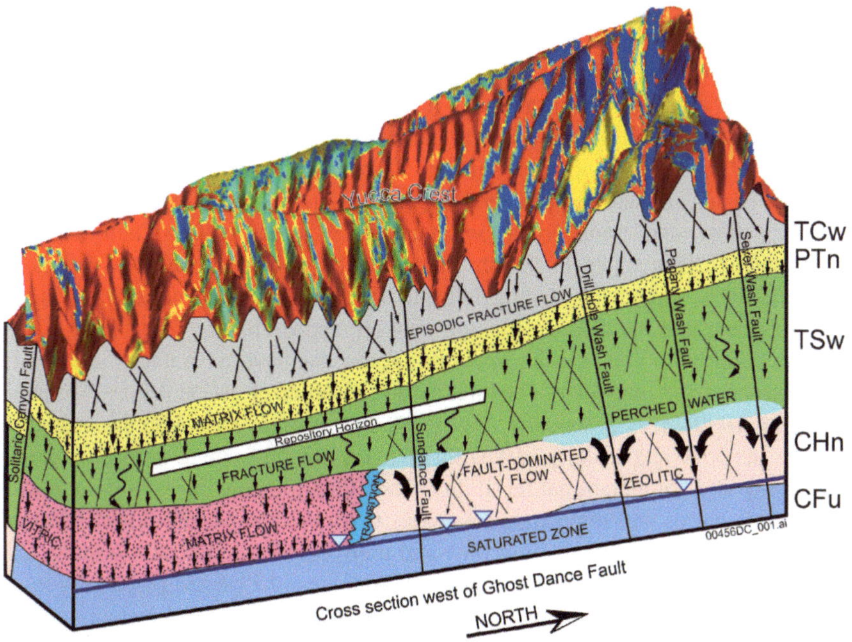

Fig. 2.15 Stratigraphic units and other features of the Yucca Mountain site

sometimes altered Calico Hills Formation (CHn); and the mostly nonwelded and altered Crater Flat (undifferentiated) Group (CFu) (Fig. 2.15). The nonwelded zones near the water table in the CHn and CFu can be subject to zeolitic alteration that reduces the matrix permeability by orders of magnitude (Bandurraga and Bodvarsson 1999). Furthermore, each stratigraphic unit is subdivided into a number of hydrogeologic layers each of which is considered to have uniform hydraulic properties. There are totally 35 hydrogeologic layers in the unsaturated zone of Yucca Mountain (Bandurraga and Bodvarsson 1999).

Welded formations are highly fractured and have low matrix permeability, while nonwelded formations have relatively small fracture densities and relatively high matrix permeability. Consequently, liquid water flow occurs mainly in the tuff matrix for nonwelded formations and in fractures for welded formations. Conceptual models of flow and transport in the unsaturated zone of Yucca Mountain can be found in Bodvarsson et al. (2000), Flint et al. (2001), and Liu (2004). For each of the 35 hydrogeologic layers, hydraulic properties include permeability and van Genuchten parameters for both fracture and matrix continua, with γ (Eq. 2.59) as an additional parameter. These properties are estimated based on small-scale property measurements and by model calibration against field data. The model calibration, also called history matching in the literature of reservoir engineering, is a procedure to adjust values of model parameters or properties such that model results match field observations. The property measurements are used to constrain the calibration results. The methodology of the model calibration and the data types used for the calibration are reported in Bandurraga and Bodvarsson (1999) and Liu et al. (1998).

Flow and transport in unsaturated fractured rocks are generally complicated because of the complexity of fracture-matrix interaction mechanisms, distinct differences in hydraulic properties between fractures and matrix, and nonlinearity involved in unsaturated flow. Several approaches are available in the literature for modeling flow and transport in unsaturated fractured rocks. When classified according to the manner in which fracture networks are treated in the model structure, these approaches mainly fall into one of the two categories: the continuum approach and the discrete fracture network approach.

Both continuum and fracture network approaches have advantages and limitations. Because the number of fractures is on the order of 10^9 at Yucca Mountain (Doughty 1999), it is practically impossible to construct and calibrate a discrete fracture network site-scale model with so many fractures in it, considering data availability and computational feasibility. Therefore, the dual-continuum approach, in which fractures and matrix are treated as two overlapping and interactive continua, has been used for modeling flow and transport within Yucca Mountain (Liu 2004). Because water flow in a fracture continuum can be treated practically in the same way as that in a porous medium, the generalized Darcy-Buckingham law that was developed for water flow in porous media can be applied to modeling unsaturated water flow in fractures. Note that the AFM was initially proposed as an empirical relation and later shown to be equivalent to the generalized Darcy-Buckingham law (Liu et al. 1998; Liu 2011a).

Gravitational fingering flow occurs at two different scales in fractures, at the single fracture scale and the fracture network scale (Fig. 2.16). The AFM was developed to deal with this fingering flow processes. The active fracture concept is based on the reasoning that, because of fingering flow, only some fractures in a connected, unsaturated fracture network contribute to liquid water flow, while other fractures are simply bypassed. The portions of the connected fractures that actively conduct water are called active fractures. We hypothesize that the number of active fractures in the unsaturated zone of Yucca Mountain is small compared to the total number of connected fractures, such that active fractures, rather than total connected fractures, must be used in numerical models for water flow. We further hypothesize that the number of active fractures within a grid block is large such that the continuum approach is still valid for describing fracture flow. These hypotheses are consistent with the consideration that fractures conducting water in the unsaturated zone of Yucca Mountain are many and highly dispersed (Liu 2004). When using the active fracture concept for modeling flow and transport in fractures, we treat active fractures as a portion of the "homogeneous" fracture continuum for a given grid block.

The portion of the active fracture continuum, f, is calculated with Eq. 2.59. It was the search for the physical basis of the AFM that led to the use of the optimality principle for the generalization of the Darcy-Buckingham law discussed in Sect. 2.4. Note that only the active fracture continuum, a portion of the total

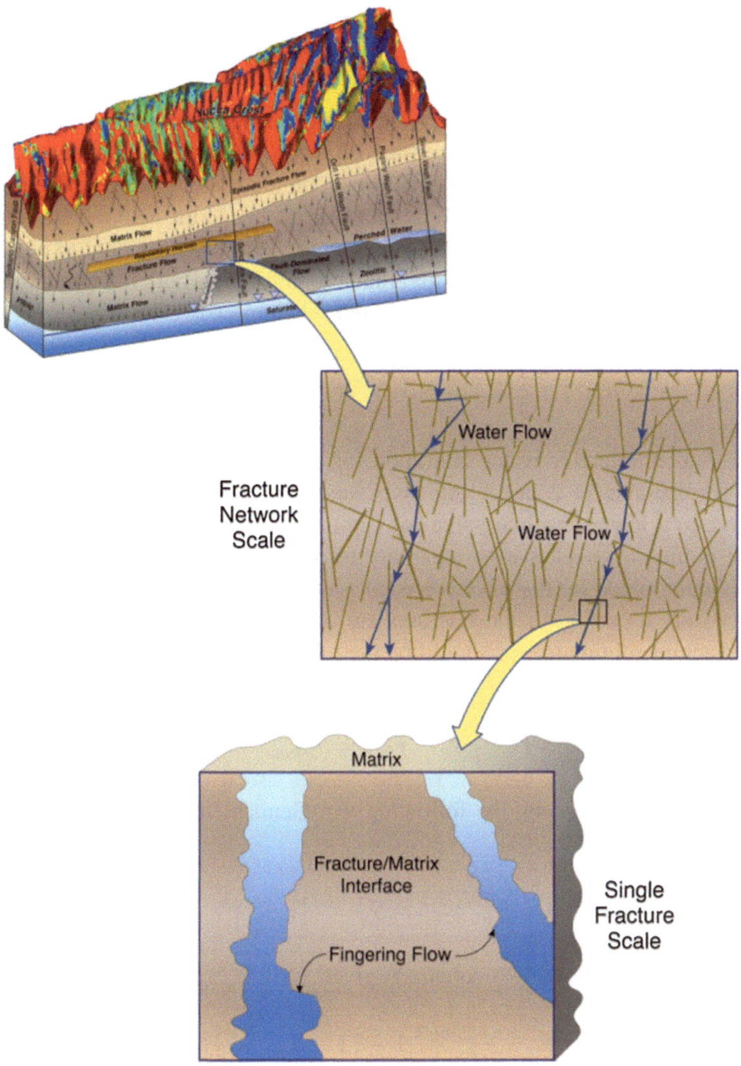

Fig. 2.16 Conceptualization of water flow within Yucca Mountain, Nevada

fracture continuum, contributes to the flow and transport in fractures and the fracture-matrix interaction. Fracture hydraulic properties should thus be defined for active fractures. Based on the water mass balance with a constant water density, the effective water saturation of active fractures, S_{af}, is related to the effective water saturation of all the connected fractures, S_{ef}, by

$$S_{af} = \frac{S_{ef}}{f} = S_{ef}^{1-\gamma} \tag{2.69}$$

Because $S_{af} \leq 1$ by definition, γ should be smaller than or equal to one.

If all connected fractures are considered to be active in conducting water, the water capillary pressure for the fracture continuum may be described by the well-known van Genuchten (1980) relation (Eq. 2.23). In the active fracture model, however, the van Genuchten capillary-pressure relation is considered to be relevant for the active fracture continuum only, rather than for the whole fracture continuum. Replacing S_e with S_{af} in the van Genuchten relation leads to

$$h(S_{ef}) = \frac{1}{\alpha}[S_{af}^{-1/m} - 1]^{1/n_{VG}} = \frac{1}{\alpha}[S_{af}^{(\gamma-1)/m} - 1]^{1/n_{VG}} \tag{2.70}$$

For a given effective water saturation for all the connected fractures, a larger γ value corresponds to a larger effective water saturation in active fractures, and therefore to a lower absolute value for capillary pressure (Fig. 2.17a).

The liquid-phase relative permeability for the active fracture continuum, k_{ar}, is directly determined by the effective water saturation of active fractures. However, because only a portion of the fractures are active, the relative permeability of the entire fracture continuum, k_r, should be the relative permeability of active fractures multiplied by f, or

$$k_r = f k_{ar} = S_{ef}^{\gamma} k_{ar} \tag{2.71}$$

where k_{ar} can be given by the following van Genuchten's (1980) relative permeability relation:

$$k_{ar} = S_{af}^{1/2}[1 - \{1 - S_{af}^{1/m}\}^m]^2 = S_{ef}^{(1-\gamma)/2}[1 - \{1 - S_{ef}^{(1-\gamma)/m}\}^m]^2 \tag{2.72}$$

Combining the above two equations gives

$$k_r = S_{ef}^{(1+\gamma)/2}[1 - \{1 - S_{ef}^{(1-\gamma)/m}\}^m]^2 \tag{2.73}$$

In general, relative permeability (k_r) is affected by γ in a complicated manner for a given S_{ef}. A larger γ value, resulting in a higher effective water saturation in active fractures (S_{ae}), gives rise to a larger value of k_{ar}. On the other hand, a larger γ value corresponds to a smaller value of f. Because the former effect is dominant, a larger γ value gives a larger relative permeability for a given effective water saturation of the fracture continuum (Fig. 2.17b).

In an unsaturated fracture network, the ratio of the fracture-matrix interface area (contributing to flow and transport between fractures and the matrix) to the total interface area (determined geometrically from the fracture network) is called the fracture-matrix interface area reduction factor by Liu et al. (1998). In the active

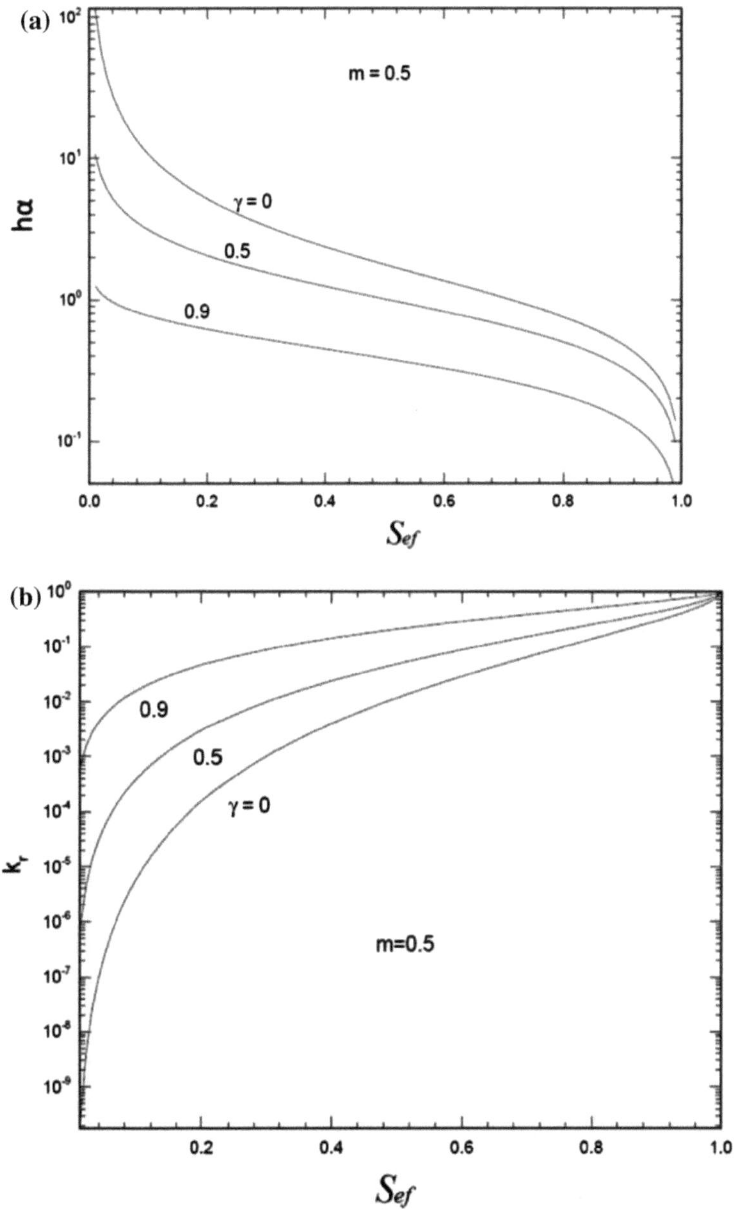

Fig. 2.17 **a** Capillary pressure curves and **b** relative permeability curves for γ = 0, 0.5 and 0.9

fracture model, this reduction factor results from three considerations. Firstly, the average interface area between mobile water in an active fracture and the surrounding matrix is smaller than the geometric interface area. Secondly, the number of active fractures is smaller than that of connected fractures. Thirdly, average active fracture spacing is much larger than that for the connected fractures; under the quasi-steady-state condition, flow and transport between fractures and their surrounding matrix is inversely proportional to the corresponding fracture spacing. Based on these considerations, Liu et al. (1998) derived an expression for the reduction factor:

$$R_{fm} \cong S_{ef}^{1+\gamma} \tag{2.74}$$

Conventionally, all the connected fractures are assumed to contribute to fracture-matrix interaction, which gives $R_{fm} = 1$. According to the above equation, the reduction factor can be significantly smaller than one as a result of fingering flow, especially when fracture water-saturation is low.

2.6.3 Verification of the AFM with Field Observations

Since the AFM was used as the base case theory for all models of flow and transport within the unsaturated zone in the Yucca Mountain Project, it has been evaluated by a number of studies. It was shown that simulation results based on the AFM match the rock-matrix water saturation and potential data collected from long boreholes at the Yucca Mountain site (e.g., Liu et al. 1998). These simulation results are also consistent with flow and tracer transport data observed from a field test (on a scale of about 20 m) performed in a densely fractured unit within Yucca Mountain (Liu et al. 2003a). This subsection focuses on evaluating the AFM with C-14 age data and fracture coating data that provide important information regarding large-scale unsaturated flow processes in fractures and fracture-matrix interaction within Yucca Mountain. Discussions in this subsection are mainly based on materials from Liu (2004) and Liu et al. (2003b).

Carbon-14 age data have been collected from different locations in the unsaturated zone of Yucca Mountain, including perched water, pore water, and gas samples (Yang 2002; Fabryka-Martin et al. 2000). They largely reflect the time for water travel from the ground surface to the locations where the data were collected. Carbon-14 data collected from perched water are not used for validating the AFM because water travel time from the ground surface to the perched-water bodies was dominated by PTn, where flow occurs mainly in the rock matrix, and thus simulated water travel time to the perched water bodies is insensitive to the AFM parameters. (Note that the AFM was developed for modeling water flow in fractures.) Pore-water carbon-14 data from various boreholes at Yucca Mountain are not considered to be representative of the pore-water residence time because of probable contamination by atmospheric $^{14}CO_2$ during drilling, resulting in apparently shorter residence times (Yang 2002; Fabryka-Martin et al. 2000). Yang (2002) reported that carbon-14 data from gas samples are most representative of in situ pore-water conditions based on two considerations. Firstly, atmospheric $^{14}CO_2$

during drilling is not expected to contaminate those gas samples collected several years after drilling. Secondly, gas phase carbon-14 age should be the same as that of the local pore water because of the rapid exchange of gas-phase CO_2 (in hours to days) with dissolved CO_2 (and HCO_3^-) in pore water. Furthermore, the amount of carbon in an aqueous-phase reservoir is hundred-times greater than carbon in the CO_2 gas-phase reservoir. Consequently, the aqueous phase will dominate the gaseous phase when exchange occurs, indicating the reasonableness of using gas phase carbon-14 age to represent the corresponding pore-water age (Yang 2002). Gas samples were collected from two kinds of boreholes: open surface-based boreholes and instrumented (closed) surface-based boreholes. The data from the latter boreholes (USW SD-12 and USW UZ-1) are the most reliable indicators of the in situ conditions, because these boreholes are not directly connected to the atmosphere such that the contamination of gas samples from atmospheric $^{14}CO_2$ was minimal (Fabryka-Martin et al. 2000). Thus, carbon-14 residence ages (Fabryka-Martin et al. 2000) calculated with the gas phase data from these two boreholes are used for validating the AFM.

The fracture-coating data are also useful for validating the AFM. These data have important implications for water flow through Yucca Mountain since they are generally a signature of historical water flow paths. The process of unsaturated-zone mineral deposition is initiated during infiltration where meteoric water interacts with materials in the soil, after which a portion may then enter the bedrock fracture network, resulting in mineral coatings on fracture surfaces. Fracture coating data were collected in an underground tunnel that was constructed for investigating field-scale flow and transport processes in the unsaturated zone of Yucca Mountain. Spatial distribution of fractures with observed coatings was used to estimate the portion of active fractures. For a given survey interval along the tunnel, characters of fractures intersecting the survey line were documented. The ratio of the number of coated fractures along the survey line within a geological unit to the number of total fractures within the same unit provides an estimate of the portion of active fractures for the unit. The estimated average portion of active fractures for the TSw is about 7.2 %. Note that fracture coatings may not precisely represent all the active flow paths in the unsaturated zone of Yucca Mountain, because not all the flow paths correspond to fracture coatings (Liu et al. 1998). Nevertheless, these values give us at least an order-of-magnitude estimate of the portion of active fractures that is about 10 %. The data also show that mineral-coating growth rate has been more or less constant with time in the unsaturated zone of Yucca Mountain, implying that the fracture network there has maintained a large degree of hydrologic stability and fracture flow paths in the deep unsaturated zone are buffered from climate-induced variations in precipitation and infiltration (Fabryka-Martin et al. 2000). If the AFM accurately represents water flow processes in the unsaturated zone of Yucca Mountain, modeling results based on the AFM should be consistent with this important observation.

The AFM-based simulation results are compared with the carbon-14 age data. One-dimensional dual-permeability numerical models are developed for vertical flow and transport along rock columns corresponding to boreholes USW SD-12 and

USW UZ-1. Calibrated rock properties reported by Liu and Ahlers (2003) are used in the models, except for γ values associated with the TSw formation (where the repository is located). The value of the AFM parameter γ for the TSw formation is varied for different simulations to check the sensitivity of simulated water travel times to this parameter within the formation, because water travel time corresponds to the carbon-14 age. The present-day infiltration rate and a constant tracer concentration are used as the top boundary conditions for flow simulations and for transport simulations. The tracer concentration is assumed to be zero initially within the fractured rocks. The dispersion process is ignored here, because previous studies indicate that dispersion has an insignificant effect on overall solute transport behavior in unsaturated fractured rocks (Bodvarsson et al. 2000; Liu et al. 2000). An effective diffusion coefficient value of 1.97E−10 m^2/s, equal to the average value of coefficients for tritiated water measured from matrix samples from the Yucca Mountain site, is employed for modeling tracer transport. TOUGH2 and T2R3D codes (Pruess 1991; Wu et al. 1996) are used for simulating steady-state water flow and tracer transport processes. There is good agreement between simulation results using the above two codes for solute transport in the unsaturated zone of Yucca Mountain, and those obtained using a particle tracking method without numerical dispersion (Liu et al. 2000), indicating that the effects of numerical dispersion are indeed insignificant for the problem under consideration. Simulated water travel times (or ages) for the rock matrix are compared with carbon-14 ages that are believed to correspond to the matrix. At a given location, the simulated water travel time is determined as the time when the matrix concentration reaches 50 % of the top-boundary concentration because it represents the average travel time for water particles from the ground surface to the location under consideration.

Figure 2.18 shows vertical distributions of simulated water travel times (ages) for different γ values. The simulated results within the TSw unit are quite sensitive to the AFM parameter γ, indicating that carbon-14 data are useful for validating the AFM and for constraining the γ values for the TSw unit. For γ values ranging from 0.2 to 0.4, simulated results approximately match the observations for the two boreholes simultaneously, although, as a result of subsurface heterogeneity, a better match is obtained for USW SD-12 than that for USW UZ-14. (The heterogeneity within each geological layer is not considered in the numerical models.) The water travel time at a given location is the summation of travel times within fractures and matrix, but is dominated by that in the matrix for the TSw unit because matrix diffusion there is a much slower process than advection in fractures. Thus, a larger γ value generally corresponds to a longer travel time, resulting from the smaller degree of matrix diffusion owing to a smaller fracture-matrix interfacial area available for mass transport between fractures and matrix (Eq. 2.74). Because of the spatial variability of the degree of matrix diffusion, the simulated water travel times and the observed ages in Fig. 2.18 are not always monotonically increasing with depth in the TSw.

There is a sharp decrease in simulated water travel times at an elevation of about 1100 m for the two boreholes (Fig. 2.18). This is because the upper portion of the

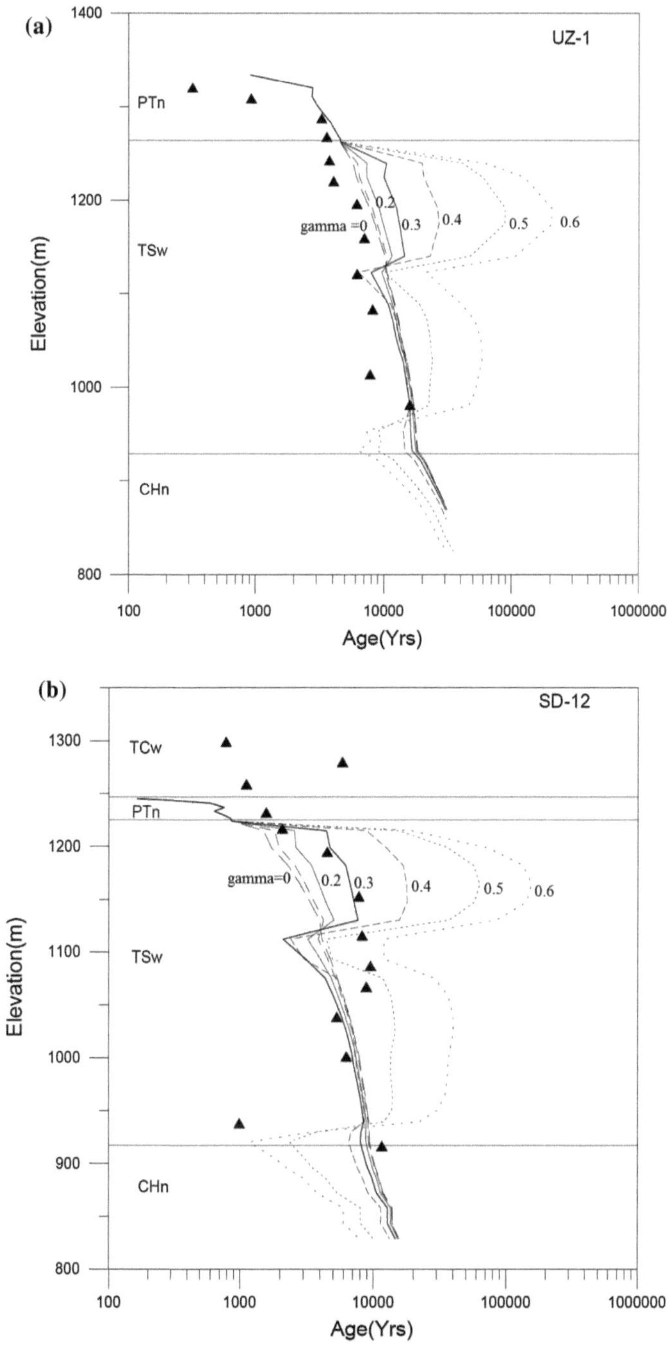

Fig. 2.18 Comparisons between simulated water travel times (ages) for rock matrix at boreholes **a** USW UZ-1 and **b** USW SD-12 and the corresponding Carbon-14 ages for several γ values (Liu 2004). The *solid triangles* are measurements and the γ interval is 0.1 for the simulated curves

TSw unit above that elevation has relatively small fracture density values and therefore corresponds to a smaller degree of matrix diffusion for a given γ value, or longer water travel times for the rock matrix. For the borehole USW UZ-1, simulated water travel time is generally longer than the observed age data at a given elevation, which may result from a subsurface heterogeneity that gives larger fracture densities at the borehole location than what are used in the numerical model. (Layer-averaged fracture properties are used in the numerical model.) In general, Fig. 2.18 indicates that simulated water travel times, based on the AFM with γ values for the TSw between 0.2 and 0.4, reasonably represent the observed carbon-14 ages.

To check the consistency of the AFM with the coating data, the one-dimensional model for borehole USW SD-12, described above, is further used for estimating the portion of the active fractures under different climate conditions. As indicated previously, the fracture coating data is the signature of water flow history in fractures. USW SD-12 is located near the middle of the underground tunnel where coating data were collected, and thus chosen here for the model study. Two infiltration rates (Flint et al. 1996), present day mean infiltration rate (3.4 mm/year) and glacial maximum infiltration rate (17.3 mm/year), are employed as the top boundary conditions. The latter infiltration rate is about five times as large as the former rate and represents the maximum infiltration rate in the past climates. Again, uniform γ distributions for the TSw formation are assumed.

Figure 2.19 shows the simulated average portion of active fractures, f, for the TSw formation as a function of γ for the two infiltration rates. The average effective saturation for the TSw formation is used to calculate f with Eq. 2.59. For γ values ranging from 0.4 to 0.2, that are similar to those used for matching the carbon-14 data, the calculated f values are about 10–40 %. Recall that the active fracture portion estimated from fracture coating data in the TSw was about 10 %. This estimate corresponds to the lower limit of the calculated f values because not all the active flow paths are associated with coatings. For example, Wang et al. (1999) found that a flow feature does not have coatings under ambient conditions in the unsaturated zone of Yucca Mountain. Since the number of active fractures increases with γ, $\gamma = 0.4$, giving rise to $f = 10$ %, may roughly represent the upper limit for

Fig. 2.19 Simulated average portion of active fractures for the TSw formation as a function of infiltration rate and γ

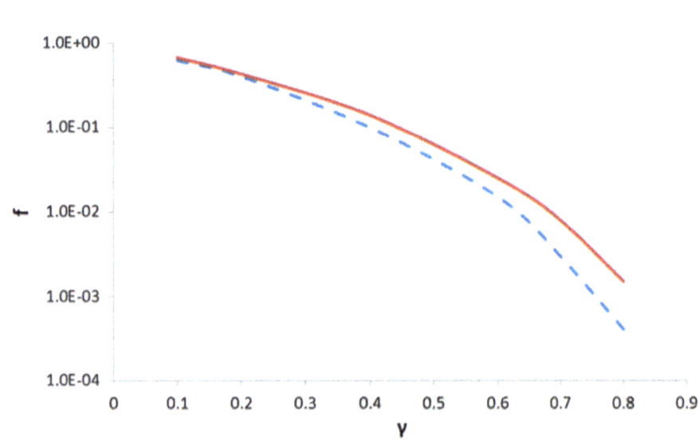

the actual γ values. For γ values less than 0.4, the calculated f values do not change significantly for the two infiltration rates (Fig. 2.19), which is consistent with the observation of the stability of flow paths over time associated with different climate conditions.

In summary, Liu et al. (2003b) and Liu (2004) showed that the AFM-based simulation results, for γ = 0.2–0.4, approximately match the observed carbon-14 age data collected from the two boreholes, and they are consistent with the portion of active fractures estimated from the fracture coating data in the TSw unit. The simulated portion of active fractures range from 10 to 40 %, while 10 % is believed to represent the lower limit for the actual portion, as previously discussed. The insensitivity of the portion of active fracture (where water flows) to infiltration rates, for γ = 0.2–0.4, is also consistent with the stability of flow paths over time that was observed from the unsaturated zone of Yucca Mountain. All these support the validity of the AFM in a practical sense.

2.6.4 Comparisons with Fracture Network Modeling Results

Başağaoğlu et al. (2009) presented a detailed study on unsaturated flow behavior in a two-dimensional fracture network (Fig. 2.20). The results are valuable for improving our understanding of unsaturated flow processes, but not really ideal for evaluating the AFM that was developed for capturing large-scale preferential flow in fractured rocks, because the corresponding fracture-network geometry (Fig. 2.20) is relatively idealized. Nevertheless, the data set is useful for demonstrating the validity of the AFM in a corroborative manner (Liu 2010).

The numerical experiments (simulations) conducted by Başağaoğlu et al. (2009) are based on the lattice-Boltzmann method. (The lattice-Boltzmann method represents fluid as an ensemble of particles that synchronously stream along bonds of a

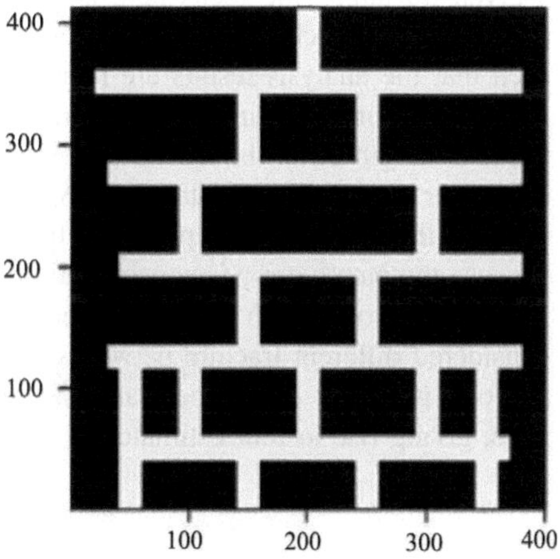

Fig. 2.20 Fracture network used by Başağaoğlu et al. (2009) for numerical experiments

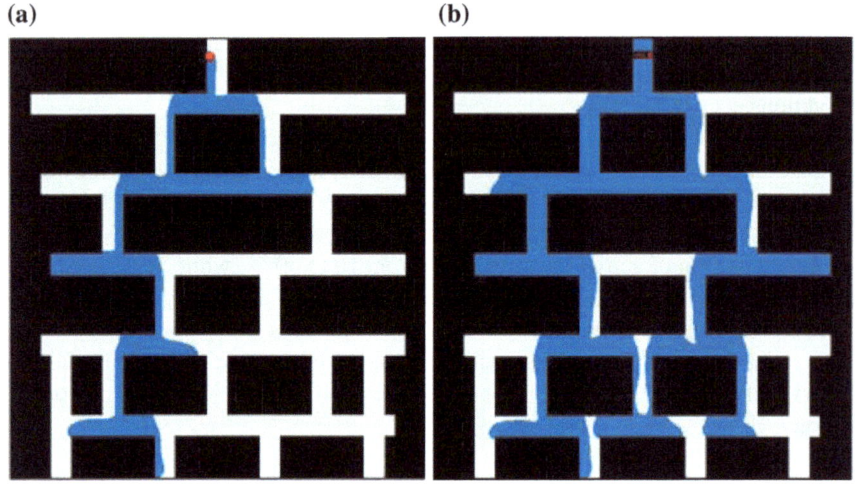

Fig. 2.21 Flow patterns for a tilt angle of 2.5°. The injection rate in (**b**) is four times as large as that in (**a**) (Başağaoğlu et al. 2009)

regular lattice and undergo mass- and momentum-conserving collisions at nodes.) The fracture network has a uniform fracture aperture and a total of 19 flow channels, consisting of 5 long horizontal and 14 short vertical channels (Fig. 2.20). Water was injected at an injection port from the topmost vertical channel at a constant rate for different tilt angles of inclination for the network. For a given tilt angle, four injection rates were used; the ratios of the first injection rate to the subsequent injection rates are 1/2, 1/3 and 1/4, respectively. Figure 2.21 shows flow patterns for a tilt angle of 2.5°; the injection rate in Fig. 2.21b is four times as large as that in Fig. 2.21a.

The unique aspect of the study by Başağaoğlu et al. (2009) is that the values for S_e and the active fracture portion f are directly available from simulation results. For a given tilt angle, the focus here is on simulation results when the flow reaches steady state or is close to it (e.g., after water reaches the bottom of the network), such that the analysis results are representative of the whole network. The active fracture portion f is estimated by the ratio of length of fractures containing water to the total length of all the fractures in the fracture network. Based on Eq. 2.59, the parameter γ can be calculated once S_e and f are known. Başağaoğlu et al. (2009) reported that the γ value depends on fracture network orientation. This is expected because the parameter γ is considered a constant for a given fracture network, and fracture networks with the same geometry but different orientations should be considered different fracture networks.

For a given tilt angle, there are four numerical experiments with different flow rates, giving rise to four estimates of parameter γ. Başağaoğlu et al. (2009) shows

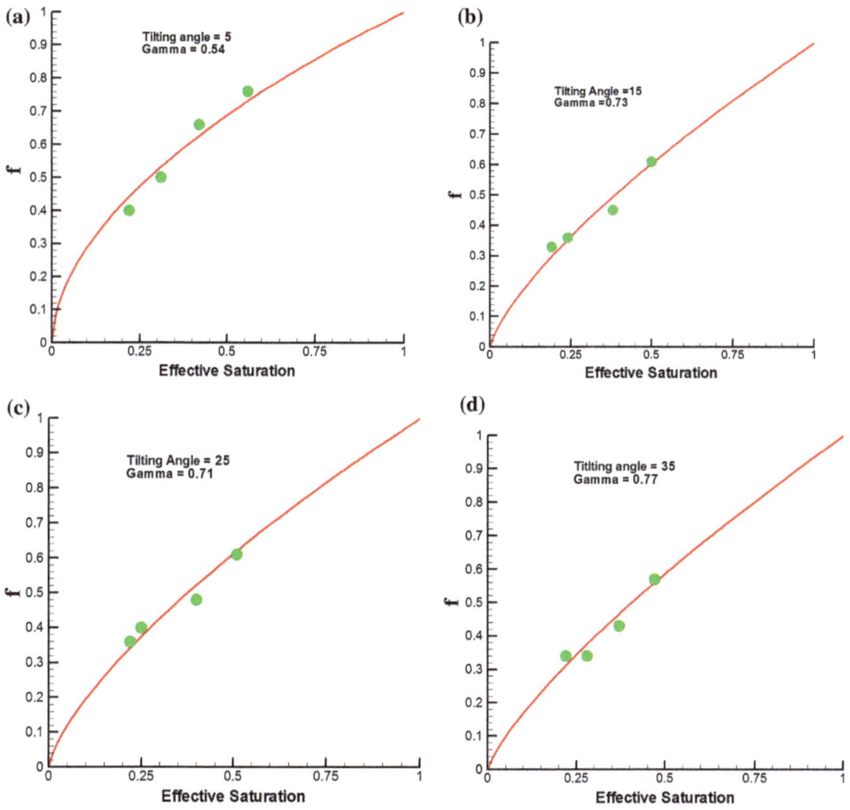

Fig. 2.22 A comparison between the theoretical relationship (Eq. 2.59) and simulation results (*data points*) from Basağaoğlu et al. (2009) for tilt angles of 5, 15, 25 and 35° [modified from (Liu 2010)]

that the γ value varies for each tilt angle, but the variation is considerably smaller than that among different tilt angles. Shown in Fig. 2.22 are comparisons between results calculated from Eq. 2.59 and those determined from Başağaoğlu et al. (2009). The averaged γ value of Başağaoğlu et al. (2009) is used for the calculation with Eq. 2.59 for a given tilt angle. Thus, no curve-fitting exercise is conducted in Fig. 2.22. Given the simplicity of the AFM and the complexity of the unsaturated flow processes in a fracture network, the AFM relationship reasonably matches the simulation results of Başağaoğlu et al. (2009), as shown in Fig. 2.22. In other words, the AFM has done a reasonable job in characterizing key macroscopic water-flow features simulated by Başağaoğlu et al. (2009).

2.7 Optimality and Surface Water Flow

The focus of this chapter is on the generalization of the Darcy-Buckingham law for water flow in unsaturated media including both soils and fractured rocks. However, the discovered power-law relationship between the conductivity and the corresponding flux seems to be general. To demonstrate this point, this section presents such a relationship for water flow on the land surface owing to rainfall events. The discussion here is based on the study of Liu (2011b).

Following Liu (2011b), we consider a landscape involving steady-state water flow and a surface evolution processes. (The latter refers to a stabilized landscape that does not change significantly with time.) This steady-state assumption has been implicitly employed in previous studies on topological structures of channel networks (Leopold and Maddock 1953; Howard 1990; Rodriguez-Iturbe et al. 1992; Rinaldo et al. 2006). (A more detailed justification of this steady-state assumption can be found in Liu (2011b).)

Based on the above simplifications, coupled water-flow (over a land surface) and surface-elevation equations can be derived from the principle that global energy expenditure rate is at a minimum. From water mass conservation and considering water density to be constant, the steady-state water flow equation is given by

$$\frac{\partial q_{s,x}}{\partial x} + \frac{\partial q_{s,y}}{\partial y} = Q_s \tag{2.75}$$

where x and y are two horizontal coordinate axes, $q_{s,x}$ and $q_{s,y}$ are the components of water fluxes (vertically-averaged water velocity multiplied by water depth) along the x and y dirctions, respectively, and Q_s is the rainfall rate. The subscript s refers to the land surface in this section.

Accordingly, the energy expenditure rate for the flow over a unit land-surface area, $\Delta_{s,E}$, can be expressed as (based on energy conservation) (Liu 2011b)

$$\Delta_{s,E} = \frac{\partial (q_{s,x}E_s)}{\partial x} + \frac{\partial (q_{s,y}E_s)}{\partial y} - Q_s E_s \tag{2.76}$$

The above equation simply states that for a given unit area, the energy expenditure rate at that location is equal to the energy carried by water flowing into the area minus the energy carried by water flowing out of the area. The rainfall is assumed to have the same energy as water at the location where the rain falls. The head, E_s (a function of x and y), refers to water energy per unit weight, including both potential (corresponding to elevation z) and kinetic energy, and is given by:

$$E_s = z + \frac{v_s^2}{2g} \tag{2.77}$$

where g again is gravitational acceleration and v_s is water velocity. Note that the second term is generally small and has been ignored in some previous studies (e.g., Howard 1990; Rinaldo et al. 2006). For completeness, this term is included here.

A combination of Eqs. 2.75 and 2.76 yields

$$\Delta_{s,E} = q_{s,x}\frac{\partial E_s}{\partial x} + q_{s,y}\frac{\partial E_s}{\partial y} \tag{2.78}$$

The water flux is considered to be given by

$$q_{s,x} = -K_s(h_s, S_{s^*})\frac{\partial E_s}{\partial x} \tag{2.79-1}$$

$$q_{s,y} = -K_s(h_s, S_{s^*})\frac{\partial E_s}{\partial y} \tag{2.79-2}$$

where

$$S_{s^*} = S_s^2 = \left(\frac{\partial E_s}{\partial x}\right)^2 + \left(\frac{\partial E_s}{\partial y}\right)^2 \tag{2.79-3}$$

In Eqs. 2.79-1 and 2.79-2, $K_s(h_s, S_*)$ is conductivity and h_s (m) is water depth. Note that in Liu (2011b), the specific formulation of conductivity is given based on Manning's equation. In this study, we assume h_s to be a function of local slope S_s only (Gupta and Waymire 1989). Many studies indicate that on average a number of hydraulic parameters can be considered as functions of local slope (Leopold and Maddock 1953). In this case, K_s is a function of S_{s*} only.

When combining Eqs. 2.78 and 2.79-1 to 2.79-3, the global energy expenditure rate throughout the water-flow domain Ω is given by

$$\iint\limits_{\Omega} \Delta_{s,E}dxdy = \iint\limits_{\Omega} (-K_s S_{s^*})dxdy \tag{2.80}$$

The optimality principle in our problem is to minimize the absolute value of the above integral. To do so, we employ the calculus of variations that seeks optimal (stationary) solutions to a functional by identifying unknown functions (Sect. 2.3). In this section, the functional corresponds to the integral defined in Eq. 2.80 and the unknown function to land-surface elevation distribution $z(x, y)$.

Furthermore, we employ the following constraint for the optimization problem (Liu 2011b):

$$\iint\limits_{\Omega} E_s dxdy = C \tag{2.81}$$

where C is a constant. Since E_s is mainly composed of potential energy z, the above equation essentially states that the average elevation throughout the model domain (or total volume of the landscape under consideration) remains unchanged, which is consistent with the steady-state assumption made in this study. It should be emphasized that the optimality principle corresponds to the minimization of the global energy expenditure rate, not the total energy within the model domain.

Based on Eqs. 2.79-1 to 2.79-3, 2.80 and 2.81, the Lagrangian for the given problem is given by

$$L_s = -K_s S_{s^*} + \lambda_1^* [S_{s*} - \left(\frac{\partial E_s}{\partial x}\right)^2 - \left(\frac{\partial E_s}{\partial y}\right)^2] + \lambda_2^* [E_s - C] \tag{2.82}$$

Note that the first term on the right hand side is from Eq. 2.80 and other terms are constraints from Eqs. 2.79-1 to 2.79-3 and 2.81, respectively. The Lagrange multipliers λ_1^* and λ_2^* are a function of location and a constant, respectively. The last term on the right hand side of Eq. 2.82 corresponds to the constraint defined in Eq. 2.81. The constraint related to water flow, Eq. 2.75, will be handled later for mathematical convenience.

The following Euler-Lagrange equation (Eq. 2.39) is used to determine an unknown function w associated with L_s defined in Eq. 2.82:

$$\frac{\partial L_s}{\partial w} - \frac{\partial}{\partial x}\left(\frac{\partial L_s}{\partial w_x}\right) - \frac{\partial}{\partial y}\left(\frac{\partial L_s}{\partial w_y}\right) = 0 \tag{2.83}$$

where w_x and w_y are partial derivatives with respect to x and y, respectively. In this study, w corresponds to S_{s*} and E_s, respectively.

Applying the Euler-Lagrange Equation (Eq. 2.83) to S_{s*} in Eq. 2.82 gives

$$\lambda_1^* = \frac{d(K_s S_{s*})}{dS_{s*}} \tag{2.84}$$

Applying the Euler-Lagrange Equation (Eq. 2.83) to E_s yields

$$\lambda_2^* + \frac{\partial}{\partial x}\left(2\lambda_1^* \frac{\partial E_s}{\partial x}\right) + \frac{\partial}{\partial y}\left(2\lambda_1^* \frac{\partial E_s}{\partial y}\right) = 0 \tag{2.85}$$

For the optimization results to be physically valid, they must satisfy the water flow equation (Eq. 2.75). A direct comparison between Eqs. 2.75 and 2.85 and consideration of Eqs. 2.79-1 to 2.79-3 reveal that Eqs. 2.75 and 2.85 are identical under the following conditions

$$\lambda_1^* = \left(\frac{\lambda_2^*}{2Q_s}\right) K_s \tag{2.86}$$

Combining Eqs. 2.84 and 2.86 gives

$$K_s \propto S_{s^*}^{-1+\frac{\lambda_2^*}{2Q_s}} = S_s^{-2+\frac{\lambda_2^*}{Q_s}} \quad (2.87)$$

From Eqs. 2.79-1 to 2.79-3 and 2.87, water flux q_s and local slope S_s have the following relation:

$$q_s \propto S_s^{-1+\frac{\lambda_2^*}{Q_s}} \quad (2.88)$$

where

$$q_s = \sqrt{q_{s,x}^2 + q_{s,y}^2} \quad (2.89)$$

The power-function relationship between water flux (or discharge) and local slope has been intensively investigated and validated in the literature (Rodriguez-Iturbe et al. 1992; Rinaldo et al. 2006; Banavar et al. 2001). Also note that the power value in the power-function relationship varies with different site conditions. However, previous studies (Leopold and Maddock 1953; Rodriguez-Iturbe et al. 1992; Rinaldo et al. 2006; Banavar et al. 2001) indicate that the averaged exponent value is about −2 in Eq. 2.88, suggesting that $\frac{\lambda_2^*}{Q_s}$ is close to −1 in an average sense.

Equations 2.87 and 2.88 lead to a relationship between the conductivity K_s (or flux divided by the energy gradient) and flux:

$$K_s \propto q_s^{\frac{2-\frac{\lambda_2^*}{Q_s}}{1-\frac{\lambda_2^*}{Q_s}}} \quad (2.90)$$

When $\frac{\lambda_2^*}{Q_s}$ is close to −1, the exponent value in the above equation is about 1.5. The equation, again, indicates that under optimal conditions, a location where a relatively large water flux occurs corresponds to a relatively small resistance (or a large conductance). As demonstrated in Sect. 2.3 and Chap. 4, the notion that the flow conductivity is a power function of flux seems to be a common rule for different macroscopic flow systems under the optimal flow conditions.

2.8 Concluding Remarks

(A) The Darcy-Buckingham law was developed based on the local-equilibrium condition that, however, does not always hold especially when fingering flow occurs. This chapter is devoted to generalizing this important law by relaxing the local-equilibrium condition. The new development is based on an optimality principle that water flow in unsaturated media is self-organized in such a way that the resistance to water flow is minimized. The key result is given as

$$k_r(h, S_*) = F(h)\left(\frac{q}{K}\right)^{a_0} \tag{2.54-2}$$

Unlike the classic Darcy-Buckingham law, the relative permeability is a function of water flux, which is a direct result of the non-equilibrium flow behavior.

(B) The generalized Darcy-Buckingham law has been shown to be consistent with laboratory observations and field data for gravitational fingering flow processes in unsaturated soils. It is also demonstrated that this generalization is the theoretical foundation for the active fracture model, the key constitutive relationship for modeling flow and transport in the unsaturated zone of Yucca Mountain that is the national high-level nuclear waste disposal site in the USA.

(C) The notion that conductivity is a power function of flux seems to be a general relationship for different flow systems. As an example, we used water flow process on the land surface to demonstrate this result. It will be further discussed in Chap. 4.

(D) While the focus of this chapter is on gravitational fingering flow in unsaturated media, the methodology is likely applicable to viscous fingering that, under certain conditions, is important for fluid flow in oil and gas reservoirs. This needs further research.

Appendix: An Alternative Derivation of Eq. 2.48 Without Using the Lagrange Multiplier

Equation 2.48 was derived in Sect. 2.4 based on the calculus of variations involving a Lagrange multiplier (Eq. 2.46). For the convenience of readers who are not familiar with the mathematical background of the Lagrange multiplier, this appendix provides an alternative derivation of Eq. 2.48 without using the Lagrange multiplier.

From Eq. 2.46, the Lagrangian (without using the Lagrange multiplier) for the given problem is

$$L = -K_{un} S_* \tag{2.A1}$$

where S_* is given as:

$$S_* \equiv \left(\frac{\partial E}{\partial x}\right)^2 + \left(\frac{\partial E}{\partial y}\right)^2 + \left(\frac{\partial E}{\partial z}\right)^2 \tag{2.44-5}$$

Equation 2.48 is then derived by solving the following Euler-Lagrangian equation with $w = E$:

$$\frac{\partial L}{\partial w} - \frac{\partial}{\partial x}\left(\frac{\partial L}{\partial w_x}\right) - \frac{\partial}{\partial y}\left(\frac{\partial L}{\partial w_y}\right) - \frac{\partial}{\partial z}\left(\frac{\partial L}{\partial w_z}\right) = 0 \tag{2.39}$$

Because $E = h + z$, we have

$$\frac{\partial L}{\partial w} = \frac{\partial L}{\partial E} = \frac{\partial L}{\partial h}\frac{\partial h}{\partial E} = \frac{\partial L}{\partial h} = -\frac{\partial (K_{un} S_*)}{\partial h} \tag{2.A2}$$

For $w_x = \frac{\partial E}{\partial x}$, we also have

$$\frac{\partial L}{\partial w_x} = -\frac{\partial (K_{un} S_*)}{\partial S_*}\frac{\partial S_*}{\partial w_x} \tag{2.A3-1}$$

$$\frac{\partial (K_{un} S_*)}{\partial S_*} = K_{un} + \frac{\partial K_{un}}{\partial (\log S_*)} \tag{2.A3-2}$$

$$\frac{\partial S_*}{\partial w_x} = 2\left(\frac{\partial E}{\partial x}\right) \tag{2.A3-3}$$

Combining Eqs. 2.A3-1 to 2A3-3 yields:

$$\frac{\partial L}{\partial w_x} = -2K_{un}\frac{\partial E}{\partial x} - 2\frac{\partial K_{un}}{\partial (\log S_*)}\frac{\partial E}{\partial x} \tag{2.A4}$$

In a similar procedure, we obtain

$$\frac{\partial L}{\partial w_y} = -2K_{un}\frac{\partial E}{\partial y} - 2\frac{\partial K_{un}}{\partial (\log S_*)}\frac{\partial E}{\partial y} \tag{2.A5}$$

$$\frac{\partial L}{\partial w_z} = -2K_{un}\frac{\partial E}{\partial z} - 2\frac{\partial K_{un}}{\partial (\log S_*)}\frac{\partial E}{\partial z} \tag{2.A6}$$

Inserting Eqs. 2.A2 and 2.A4–2.A6 into Eq. 2.39 and using the following form of water mass balance equation;

$$\frac{\partial\left(K_{un}\frac{\partial E}{\partial x}\right)}{\partial x}+\frac{\partial\left(K_{un}\frac{\partial E}{\partial y}\right)}{\partial y}+\frac{\partial\left(K_{un}\frac{\partial E}{\partial z}\right)}{\partial z}=0 \tag{2.A7}$$

we finally obtain

$$\frac{\partial\left(\frac{\partial K_{un}}{\partial(\log S_*)}\frac{\partial E}{\partial x}\right)}{\partial x}+\frac{\partial\left(\frac{\partial K_{un}}{\partial(\log S_*)}\frac{\partial E}{\partial y}\right)}{\partial y}+\frac{\partial\left(\frac{\partial K_{un}}{\partial(\log S_*)}\frac{\partial E}{\partial z}\right)}{\partial z}=\frac{S_*}{2}\frac{\partial K_{un}}{\partial h} \tag{2.48}$$

References

Banavar JR, Colaiori F, Flammini A (2001) Scaling, optimality and landscape evolution. J Stat Phys 104(1/2):1–48
Bandurraga TM, Bodvarsson GS (1999) Calibrating hydrogeologic parameters for the 3-d site-scale unsaturated zone model of Yucca Mountain, Nevada. J Contam Hydrol 38(1–3):25–46
Başağaoğlu H, Succi S, Manepally C et al (2009) Sensitivity of the active fracture model parameter to fracture network orientation and injection scenarios. Hydrogeol J 17(6):1347–1358
Bejan A (2000) Shape and structure, from engineering to nature. Cambridge University Press, New York
Bejan A, Tondeur D (1998) Equipartition, optimal allocation and the constructal approach to predicting organization in nature. Revue Generale Thermique 37:165–180
Bodvarsson GS, Liu HH, Ahlers R et al (2000) Parameterization and upscaling in modeling flow and transport at Yucca Mountain. In: Conceptual models of unsaturated flow in fractured rocks. National Academy Press, Washington DC
Brooks RH, Corey AT (1964) Hydraulic properties of porous media. Hydrology Paper No. 3, Colorado State University, Colorado
Buckingham E (1907) Studies on the movement of soil moisture. Bulletin 38, USDA Bureau of Soils, Washington DC
Buckingham E (1914) On physically similar systems: illustrations of the use of dimensional equations. Phys Rev 4(4):345–376
Burdine NT (1952) Relative permeability calculations from pore-size distribution data. Petr Trans Am Inst Mining Metall Eng 198:71–77
Doughty C (1999) Investigation of conceptual and numerical approaches for evaluating moisture, gas, chemical, and heat transport in fractured unsaturated rock. J Contam Hydrol 38:69–106
Eagleson PS (2002) Ecohydrology: Darwinian expression of vegetation form and function. Cambridge University Press, New York
Fabryka-Martin J, Meijer A, Marshall B (2000) Analysis of geochemical data for the unsaturated zone. Rep ANL-NBS-HS-000017, Yucca Mountain Project
Feder J (1988) Fractals. Plenum, New York
Flint LA, Flint LE, Bodvarsson GS (2001) Evolution of the conceptual model of unsaturated zone hydrology at Yucca Mountain, Nevada. J Hydrol 247:1–30
Flint AL, Hevesi JA, Flint EL (1996) Conceptual and numerical model of infiltration for the Yucca Mountain area, Nevada. Water Resources Investigation Report-96, US Geologic Survey, Denver, CO

Garrick BJ, Di Bella CAW (2014) Technical advances for geological disposal of high activity waste. The bridge. Nat Acad Eng 44(3):50–57
Glass RJ, Steenhuis TS, Parlarge JY (1988) Wetting front instability as a rapid and far-reaching hydrologic process in the vadose zone. J Contam Hydrol 3:207–226
Gupta VK, Waymire E (1989) Statistical self-similarity in river networks parameterized by elevation. Water Resour Res. doi:10.1029/WR025i003p00463
Heylighen F (2008) Complexity and self-organization. In: Bates MJ, Maack MN (eds) Encyclopedia of library and information sciences. Taylor & Francis, New York
Howard AD (1990) Theoretical model of optimal drainage networks. Water Resour Res 26 (9):2107–2117
Kleidon A (2009) Nonequilibrium thermodynamics and maximum entropy production in the Earth system: application and implications. Naturwissenschaften 96:653–677
Leopold LB, Langbein WB (1962) The concept of entropy in landscape evolution. Professional Paper 500-A, US Geological Survey
Leopold LB, Maddock T (1953) The hydraulic geometry of stream channels and some physiographic implications. Professional Paper 252, US Geological Survey
Liu HH (2004) Conceptual model and numerical approaches for unsaturated zone flow and transport. Report MDL-NBS-HS-000005, Yucca Mountain Project, US Department of Energy
Liu HH (2010) Comment on "Sensitivity of the active fracture model parameter to fracture network orientation and injection scenarios". Hydrogeol J 18(2):535–537
Liu HH (2011a) A conductivity relationship for steady-state unsaturated flow processes under optimal flow conditions. Vadose Zone J. doi:10.2136/vzj2010.0118
Liu HH (2011b) A note on equations for steady-state optimal landscapes. Geophys Res Lett. doi:10.1029/2011GL047619
Liu HH (2011c) Impact of climate change on groundwater recharge in dry area: an ecohydrology approach. J Hydrol 407:175–183
Liu HH (2014) A thermodynamic hypothesis regarding optimality principles for flow processes in geosystems. Chin Sci Bull 59(16):1880–1884
Liu HH, Ahlers CF (2003) Calibrated properties model. Report MDL-NBS-HS-000003 (Rev. 01), Yucca Mountain Project, US Department of Energy
Liu HH, Bodvarsson GS, Pan L (2000) Determination of particle transfer in random walk particle methods for fractured porous media. Water Resour Res 36:707–713
Liu HH, Doughty C, Bodvarsson GS (1998) An active fracture model for unsaturated flow and transport in fractured rocks. Water Resour Res 34:2633–2646
Liu HH, Haukwa CB, Ahlers CF et al (2003a) Modeling flow and transport in unsaturated fractured rock: an evaluation of the continuum approach. J Contam Hydrol 62–63:173–188
Liu HH, Zhang G, Bodvarsson GS (2003b) The active fracture model: its relation to fractal flow behavior and a further evaluation using field observations. Vadose Zone J 2:259–269
Liu HH, Zhang RD, Bodvarsson GS (2005) An active region model for capturing fractal flow patterns in unsaturated soils: model development. J Contam Hydrol 80(1–2):18–30
Montazer P, Wilson WE (1984) Conceptual hydrologic model of flow in the unsaturated zone, Yucca Mountain, Nevada. Water-Resources Investigations Report 84-4345, US Geological Survey
Mualem Y (1976) A new model of predicting the hydraulic conductivity of unsaturated porous media. Water Resour Res 12:513–522
Mualem Y, Dagan G (1978) Hydraulic conductivity of soil: unified approach to the statistical methods. Soil Sci Soc Am J 42(3):392-392
Muskat M, Meres MW (1936) The flow of heterogeneous fluids through porous media. Physics 7 (9):346–363
Nieven RK (2010) Minimization of a free-energy-like potential for non-equilibrium flow systems at steady state. Phil Trans R Soc. doi:10.1098/rstb.2009.0296
Nimmo JR, Landa ER (2005) The soil physics contributions of Edgar Buckingham. Soil Sci Am J 69:328–342

Pruess K (1991) TOUGH2-A general purpose numerical simulator for multiphase fluid and heat flow. Rep LBNL-29400, Lawrence Berkeley National Laboratory
Richards LA (1931) Capillary conduction of liquids through porous mediums. Physics 1(5):318–333
Rinaldo A, Banavar JR, Maritan A (2006) Trees, networks and hydrology. Water Resour Res. doi:10.1029/2005WR004108
Rinaldo A, Rodriguez-Iturbe I, Rigon A (1992) Minimum energy and fractal structures of drainage networks. Water Resour Res 28:2183–2191
Rodriguez-Iturbe I, Rinaldo A, Rigon A (1992) Energy dissipation, runoff production and the three-dimensional structure of river basins. Water Resour Res 28(4):1095–1103
Sheng F, Wang K, Zhang RD (2009) Characterizing soil preferential flow using iodine–starch staining experiments and the active region model. J Hydrol 367(1–2):115–124
Tondeur D, Kvaalen E (1987) Equipartition of entropy production: an optimality criterion for transfer and separation processes. Ind Eng Chem Res 26:56–65
US DOE (2002) Yucca Mountain Science and Engineering Report. Report DOE/RW-0539-1, US Department of Energy
van Genuchten MTh (1980) A closed-form equation for predicting the hydraulic conductivity of unsaturated soils. Soil Sci Soc Am J 44:892–898
Wang JSY, Trautz RC, Cook PJ et al (1999) Field tests and model analyses of seepage into drift. J Contam Hydrol 38:232–347
Wang K, Zhang R, Yasuda H (2006) Characterizing heterogeneity of soil water flow by dye infiltration experiments. J Hydrol 328:559–571
Wang Z, Feyen J, Elrick DE (1998) Prediction of fingering in porous media. Water Resour Res 34:2183–2190
Weinstock R (1974) Calculus of variations with applications to physics and engineering. Dover Publications Inc., New York
Wu YS, Ahlers CF, Fraser P (1996) Software qualification of selected TOUGH2 modules. Report LBNL-39490, Lawrence Berkeley National Laboratory
Wyckoff RD, Botset HG (1936) The flow of gas-liquid mixtures through unconsolidated sands. Physics 7(9):325–345
Yang IC (2002) Percolation flux and transport velocity in the unsaturated zone, Yucca Mountain, Nevada. Appl Geochem 17:807–817
Yasuda H, Berndtsson R, Persson H (2001) Characterizing preferential transport during flood irrigation of a heavy clay soil using the dye Vitasyn Blua. Geoderma 100:49–66

Chapter 3
Two-Part Hooke Model (TPHM): Theory, Validation and Applications

Abstract Hooke's law was discovered by Robert Hooke (1635–1703) and states that during elastic deformation of a solid body the strain is linearly proportional to the stress. While Hooke's law was initially developed based on experimental observations for relatively homogeneous materials, natural rocks generally contain small-scale deformation heterogeneities (such as the existence of micro-cracks). To address this issue, this chapter introduces the two-part Hooke model that was developed by conceptually dividing a natural rock into hard and soft parts that are subject to different degrees of mechanical deformation under a given stress. Remarkable consistency between the model and observations from different sources, for both rock matrix and fractures, has been achieved. The usefulness of the model in dealing with several engineering problems is also demonstrated, including coal-permeability changes caused by swelling under different stress conditions, coupled hydro-mechanical processes associated with CO_2 geological sequestration in a fractured formation, and evolution of the excavation damage zone related to nuclear waste disposal in a shale formation.

Subsurface fluid flow is often coupled with mechanical deformation of subsurface media where fluid flow occurs. In many cases, this coupling can play a dominant role. Hooke's Law is the most fundamental law governing elastic deformation of solids. However, a natural rock has unique features compared with other solids such that elastic mechanical deformation of the natural rock does not follow exactly the traditional Hooke's Law; the related mechanical properties, unlike those assumed in the Hooke's Law, are not constant under certain conditions. This chapter briefly reviews Hooke's life and his law, and presents our newly developed two-part Hooke model (TPHM) that incorporates small-scale mechanical heterogeneity of natural rocks. The validation and applications of the TPHM are also discussed. Note that definition and physical meaning of a symbol (denoting a variable or function) in this chapter (that is closely related to rock mechanics) may be different from the same symbol in the previous chapters unless the same physical meaning is explicitly indicated. This will allow this chapter to follow the conventional usage of

H.-H. Liu, *Fluid Flow in the Subsurface*, Theory and Applications of Transport in Porous Media 28, DOI 10.1007/978-3-319-43449-0_3

symbols in the literature of rock mechanics. The discussions in this chapter are mainly based on the materials from Liu et al. (2009, 2011, 2013), Zhao and Liu (2012), Liu and Rutqvist (2010, 2013) and Li et al. (2014).

3.1 Robert Hooke and His Law for Elastic Deformation

Hooke's Law was discovered by English scientist Robert Hooke (1635–1703). It states that the amount by which a material (e.g., rock) body is deformed (the strain) is linearly related to the force (stress) causing the deformation. While Robert Hooke is now primarily remembered for Hooke's Law, he was widely involved in science and had many other achievements. For example, he coined the word "cell" that is one of the most commonly used terms in biology. Robert Hooke is a very controversial and possibly underestimated figure in the science history, largely because of his involvement in several famous disputes of priorities of scientific discoveries. This section does not intend to give a comprehensive overview of Hooke's life and all his major scientific achievements, but focuses on those that may be relevant to the discovery of Hooke's Law. Readers who wish to know more about Hooke are encouraged to read his two biographies by Jardine (2004) and Inwood (2005). There is also a relatively comprehensive description of Hooke's life and scientific contributions in Wikipedia's item "Robert Hooke" (http://en.wikipedia.org/wiki/Robert_Hooke).

Robert Hooke was born in 1635. As a youth, Hooke already demonstrated extraordinary gifts to make precise measurement and to do complex mechanical works. It was reported that Hooke dismantled a brass clock and built a wooden replica that seemed to work well enough. Hooke entered Westminister School in London in about 1648 and then in 1653 went to Oxford as a chorister at Christ Church. Hooke himself characterized his Oxford days as a source of his lifelong passion for science and the friends he made there were important to him throughout his career. Some of his friends went on to form the nucleus of the Royal Society. From 1655 to 1662, Hooke worked with Robert Boyle, a professor at Oxford, as his laboratory assistant. He built up air pumps for Boyle that were believed to help discover the Boyle's gas law that describes how the pressure of a gas tends to decrease as the volume of a gas increases for a given temperature. One year later after the Royal Society was founded in 1690, Hooke was appointed as the Curator of the society, with a responsibility to demonstrate experiments from his own methods or at the suggestion of members. He then became a fellow of Royal Society and Professor of Geometry at Oxford in 1664. He received the degree of "Doctor in Physics" in 1691. Hooke was a very hard working and productive person. He had never been married and died in London in 1703.

Hooke discovered Hooke's Law in 1676 and announced his discovery in the anagram "ceiiinosssttuv". During that time, anagram was sometimes used by scientists to establish priority for a discovery without revealing details. In 1968, Hooke published the solution to the anagram as "Ut tension, sic vis" meaning "As the tension, so the force". A current statement of Hooke's Law is given in the beginning of this section. In his presentation of Hooke's Law, Hooke illustrated the law by discussing several different experimental situations, including loading a wire, a spiral spring, and a watch spring (Moyer 1977). As one of the most important physics laws, Hooke's Law gave the birth to quantitative solid mechanics and can be found in any standard high-school textbook including mechanics.

The dispute between Robert Hooke and Isaac Newton over credit for the work on gravitation has been well known in the science history (Jardine 2004; Inwood 2005). In 1686, Newton presented his book "Principia" to the Royal Society that documented the universal gravitational law. Hooke then claimed that Newton got the idea initially from him (through letter exchanges) that gravitation is inversely proportional to square of distance between enters of the two bodies, but accepted that the associated mathematical developments and demonstrations were wholly Newton's. There are different versions of what really happened. It was told that when Newton famously said that he saw farther than others because he stood on the shoulders of giants, what Newton really meant was that his discovery had nothing to do with Hooke who, in real life, was a very short person (Crease 2010). However, partially as a result of the dispute, Newton did acknowledge in all editions of "Principia" that Hooke, along with two other scientists, separately appreciated the inverse square law in the solar system. One year later after Hooke died, Newton became the president of the Royal Society in 1704 and did much to obscure Hooke, including destroying his only known portrait. Nevertheless, Robert Hooke has now been recognized as one of the most important scientists of his age.

3.2 Two-Part Hooke's Model

Mechanical deformation processes in porous and fractured rocks are often coupled with thermal and hydrological processes. These coupled processes are important in a number of areas, including geothermal energy development, oil and gas extraction, nuclear waste geological disposal, geological sequestration of carbon dioxide, and deep well injection of liquid and solid wastes. The stress-strain relationship is the fundamental requirement for modeling mechanical deformation and the associated coupled processes in porous and fractured rocks and has been generally described by Hooke's Law for elastic mechanical deformation. However, the current application of the Hooke's Law to porous and fractured rocks is not without questions. Strictly speaking, the proportionality in the observed stress-strain relationship should be constant for elastic deformation if the current application of Hooke's Law is perfectly valid, simply because according to Hooke's Law, the strain is linearly related to the stress. However, a number of studies have already indicated that the proportionality is not always constant, but rather stress-dependent in many cases (e.g., Mavko and Nur 1978; Zimmerman 1991; Al-Tahini and Abousleiman 2010). Several researchers (e.g., Walsh 1965a, b; Nur 1971) have made efforts to relate this stress-dependent behavior to the microstructures of "cracks" in porous rock. An excellent review of these efforts is presented in a chapter entitled *Micromechanical Models* in Jaeger et al. (2007). It is generally difficult to characterize small-scale structures accurately and then relate their properties, in a relatively straightforward and simple way, to large-scale mechanical properties that are of practical interest. What is more desirable for practical applications is to have a macroscopic-scale theory that does not rely on the detailed description of small-scale structures, and that can physically incorporate the stress-dependent behavior of relevant mechanical properties. This section presents a theory of this kind, the two-part Hooke model (TPHM) proposed by Liu et al. (2009). The TPHM-based constitutive relationships between stress and hydraulic/mechanic properties are also discussed.

3.2.1 TPHM for Isotropic Stress Condition

The TPHM is based on the two key arguments. The first one is that an appropriately defined strain, other than the one commonly used in the rock mechanics literature, should be used in Hooke's Law. The second one is that Hooke discovered Hooke's Law for homogeneous systems (Sect. 3.1), while a natural rock exhibits a large degree of mechanical-property heterogeneity. For example, micro-cracks have dramatically different mechanical properties than the rest of the rock mass and at the same time could dominate fluid flow process within the rock body. They should be treated differently from the rest of rock mass on the continuum scale. For simplicity, we consider the relationship here for the volumetric strain to demonstrate these arguments, although our results can be extended to other types of strains. This subsection is based on the material from Liu et al. (2009).

Under the isotropic stress (or hydrostatic stress) condition, Hooke's Law describes the elastic volumetric deformation of a homogeneous and isotropic subsurface material by

$$d\sigma = K d\varepsilon_{v,t} \tag{3.1}$$

where σ is the hydrostatic (effective) stress that is positive in the compressive direction, K is bulk modulus, and $\varepsilon_{v,t}$ is the natural or true volumetric strain defined by (Freed 1995)

$$d\varepsilon_{v,t} = -\frac{dV}{V} \tag{3.2}$$

where V is the total volume of subsurface material under the current state of stress. In Eqs. 3.1 and 3.2, the positive strain corresponds to a decrease in the volume. The first hypothesis of the TPHM is that Hooke's Law holds for natural strains. Freed (1995) provided a historical review of the development of the concept of natural strain in the literature of material science and argued that the natural strain should be used for accurately describing material deformation.

In previous studies (e.g., Jaeger et al. 2007), the following engineering strain $(\varepsilon_{v,e})$ is often used when applying Hooke's Law:

$$d\varepsilon_{v,e} = -\frac{dV}{V_0} \tag{3.3}$$

where V_0 is the bulk volume under zero stress.

Inserting Eq. 3.3 into Hooke's Law (Eq. 3.1) and performing the integration under the condition that $V = V_0$ for $\sigma = 0$, one obtains

$$V = V_0\left(1 - \frac{\sigma}{K}\right) \tag{3.4}$$

Similarly, the use of natural strain in Hooke's Law (Eq. 3.2) yields

$$V = V_0\exp\left(-\frac{\sigma}{K}\right) \tag{3.5}$$

Equations 3.4 and 3.5 are nearly identical for small values of $\frac{\sigma}{K}$ (or strain), because $\exp\left(-\frac{\sigma}{K}\right) \rightarrow 1 - \frac{\sigma}{K}$ for $\frac{\sigma}{K} \rightarrow 0$.

In the literature of rock mechanics, the engineering strain has been exclusively used; it is generally believed that the elastic strain is small for most practical applications. A porous and fractured rock, however, differs from purely solid materials in that it is inherently heterogeneous and includes both solid phase and pores (and/or fractures) with a variety of geometric shapes that exhibit a large degree of variation in mechanical properties. While the bulk elastic strain may be indeed small in most of the rock for stress changes of practical interest, the strain can be considerably larger within some portions of a rock body. For example, some pores (or fractures) in a rock can be subject to significant deformation, and even further, they can be completely closed under a certain range of stress changes encountered in practice. For these pores, the (engineering) strain is not small, but could be on the order of one. An accurate description of the deformation of this portion of the rock is important for coupled mechanical and hydrological processes, because fluid flow occurs in pores and fractures.

To deal with this issue, the TPHM conceptualizes a heterogeneous rock to have two parts, and hypothesizes that one part (a portion of pore volume or fracture apertures) needs to be described with natural-strain-based Hooke's Law, and the other part can be adequately described with engineering-strain-based Hooke's Law because the deformation is small for this part. For simplicity, the first part is called "soft" part and the other called "hard" part. This conceptualization can be represented by a hypothesized composite spring system shown in Fig. 3.1. These two springs are subject to the same stress, but their deformations are governed by different varieties of Hooke's Law. In his study on anisotropy of pore-fluid

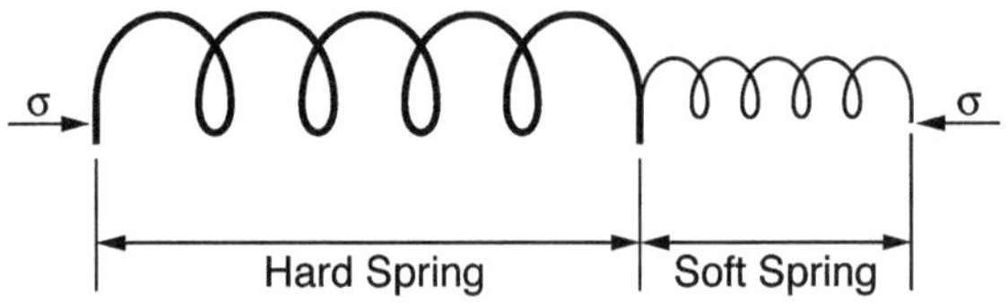

Fig. 3.1 A composite spring system consisting of two springs (Liu et al. 2009) (reproduced by permission of Elsevier). The hard and soft springs follow engineering-strain-based and natural-strain-based Hooke's Law, respectively

enhanced shear modulus, Berryman (2006) also divided a poroelastic medium into "hard" and "soft" portions for the similar purpose. Mavko and Jizba (1991) considered rock porosity to consist of a soft part and a stiff part in investigating grain-scale fluid effects on velocity dispersion in rocks. The "two-part" conceptualization of the TPHM is generally consistent with these previous studies. In this chapter, we use subscripts 0, e, and t to denote the unstressed state, the hard part and the soft part, respectively, for a rock body. Then we have

$$V_0 = V_{0,e} + V_{0,t} \tag{3.6}$$

and

$$dV = dV_e + dV_t \tag{3.7}$$

Applying Eqs. 3.4 and 3.5 to rock volumes V_e and V_t, respectively, in Eq. 3.7 yields

$$-\frac{dV}{V_0} = \gamma_e \frac{d\sigma}{K_e} + \gamma_t \exp\left(-\frac{\sigma}{K_t}\right)\frac{d\sigma}{K_t} \tag{3.8}$$

$$\gamma_t = \frac{V_{0,t}}{V_0} \tag{3.9}$$

$$\gamma_e = 1 - \gamma_t \tag{3.10}$$

where K_e and K_t are the bulk moduli for the hard and soft parts, respectively, and γ_t and γ_e are volumetric fractions of the hard and soft parts, respectively, under the unstressed condition. Equations 3.8–3.10 together comprise the proposed stress-strain relationship, or the TPHM, in terms of volumetric strain (Liu et al. 2009).

We should emphasize that the TPHM mathematically holds only when the soft and hard parts have dramatically different bulk moduli such that variations of mechanical properties within each part is much smaller than the corresponding differences between the two parts and therefore can be ignored when calculating the bulk deformation from the two parts. It is clearly the case for natural rocks, as demonstrated in the sections to be followed in this chapter.

The TPHM is a macroscopic-scale approximation that uses natural-strain-based Hooke's Law to describe nonlinear deformation behavior of a fraction of pore volume (consisting of a collection of pores with a variety of geometry) subject to larger deformations. This nonlinear deformation could result from combining effects of non-uniform pore size distributions and pore-geometry heterogeneity (Jaeger et al. 2007). A rough fracture can also be considered a collection of pores with different sizes and geometries for the purpose of deformation calculations. The validity of this approximation will be evaluated in the following sections.

3.2.2 TPHM-Based Constitutive Relationships for Isotropic Stress Condition

In this subsection, we use the TPHM to derive constitutive relationships between stress and mechanical or hydraulic properties for porous rock subject to elastic deformation under isotropic stress conditions. The derived relationships are also compared with the corresponding empirical expressions and experimental data, as an effort to verify the TPHM. This subsection is based on the materials from Liu et al. (2009) and Zheng et al. (2015).

3.2.2.1 Bulk Rock Compressibility

Bulk rock compressibility is a measure of the capability for a rock body to deform when stress is changed under a constant pore-pressure condition. Mathematically, the bulk rock compressibility is expressed as (Jaeger et al. 2007):

$$C_{bc} = -\frac{1}{V_0}\frac{\partial V}{\partial \sigma} \tag{3.11}$$

Substituting Eq. 3.8 into Eq. 3.11 yields

$$C_{bc} = \frac{\gamma_e}{K_e} + \frac{\gamma_t}{K_t}\exp\left(-\frac{\sigma}{K_t}\right) \tag{3.12}$$

The derived stress-compressibility relation consists of two terms, with the first term being a constant and the second one being an exponential function. This is consistent with an observation that rock compressibility data have often been empirically fitted to exponentially decreasing functions of the form (e.g., Zimmerman 1991; Jaeger et al. 2007; Wyble 1958):

$$C_{bc} = C_{bc}^{\infty} + \left(C_{bc}^{init} - C_{bc}^{\infty}\right)\exp\left(-\frac{\sigma}{P^*}\right) \tag{3.13}$$

where the superscript *init* denotes the initial (zero stress) value, the superscript ∞ denotes the value at high stress (or $\frac{\sigma}{P*} \rightarrow \infty$), and $P*$ is a parameter that can be considered as a characteristic stress. Based on certain assumptions, Jaeger et al. (2007) rewrote the above equation as

$$C_{bc} = C_{bc}^{\infty} + \frac{\phi_{crack}}{P^*}\exp\left(-\frac{\sigma}{P^*}\right) \tag{3.14}$$

where ϕ_{crack} is the porosity of crack-like voids in a porous rock sample. These voids are believed to be responsible for observed nonlinear deformation (Jaeger et al. 2007). The measured compressibility data of three consolidated sandstones, Boise, Berea, and Bandera, were fitted to functions of the form of Eq. 3.13 by Zimmerman (1991). We refer the readers to Zimmerman (1991) for the details of his curve-fitted results.

Several interesting observations can be made when comparing the derived expression for rock compressibility (Eq. 3.12) with the empirical relation given by Eq. 3.13 or Eq. 3.14. Firstly, our theoretical result is consistent with the empirical relations and the related experimental data used to develop the relation, as evidenced by the fact that the functional forms of Eqs. 3.12 and 3.14 are essentially identical. Secondly, the curve-fitted results of Zimmerman (1991) give ranges of γ_t (defined in Eq. 3.12 and equivalent to ϕ_{crack} in Eq. 3.14) between 0.2 and 0.5 % for the three sandstones under consideration, suggesting that the so-called soft part of the rock body is only a small percentage of pore volume. Note that a typical porosity value for a sandstone rock is between 10 and 20 %. This observation will be further confirmed by more results to be discussed later in this chapter. Thirdly, the values for K_t for the three sandstones (4.74–8.33 MPa) are significantly smaller (by three orders of magnitude) than those for K_e (9.5–12.2 GPa). In other words, the soft part is indeed much "softer" than the rest of the rock body. The observed significant difference between K_t and K_e supports the argument (used in developing the TPHM) that variations of mechanical properties within each part are much smaller than the corresponding differences between the two parts. Note that the second term on the right hand side of Eq. 3.12 is not necessarily smaller than the first one, especially for low stress states, although γ_t is generally small, as demonstrated in Fig. 3.2.

A relationship between bulk compressibility and stress was also proposed by Shapiro and his coworkers (Shapiro 2003; Shapiro and Kaslow 2005; Becker et al. 2007). Their studies are based on a key assumption that the compressibility is a linear function of crack porosity and change in the stiff porosity that is similar to rock porosity of the hard part in this study. However, the TPHM differs from their work in several important aspects. Firstly, the TPHM is based on the natural-strain-based Hooke's Law, which is fundamentally different from the physical origin of Shapiro and his coworkers. Secondly, their theory is limited to rocks with moderate or small

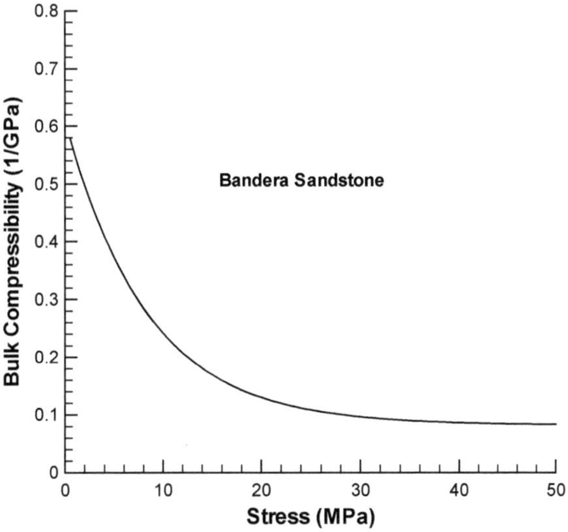

Fig. 3.2 Bulk compressibility as a function of stress calculated from Eq. 3.14 using the fitted parameter values (C_{bc}^{∞} = 0.082 GPa^{-1}, P^* = 8.33 MPa and $\phi_{crack} = 0.0044$) (Liu et al. 2009) (reproduced by permission of Elsevier)

porosity, on the order of 10 % or less (Shapiro and Kaslow 2005). As evidenced by the derivation procedures in this chapter, the TPHM-based results are not subject to this limitation. Finally, Shapiro and Kaslow's (2005) theory is valid only when their compliant (or crack) porosity is a very small portion of total porosity. The TPHM does not suffer from this limitation either, largely because the TPHM has a different physical origin. It can be applied to cases in which soft porosity is large. For example, based on the TPHM, Liu et al. (2009) successfully derived a relationship between stress and fracture aperture. Unlike the "soft" part of porous rock, the "soft" part in a fracture corresponds to a much larger portion of fracture voids than the hard part (Liu et al. 2009), as will be discussed in Sect. 3.3.

3.2.2.2 Pore Compressibility

The pore compressibility refers to relative change in pore volume per unit change in stress under a condition of constant pore pressure. Mathematically, this compressibility can be expressed by

$$C_{pc} = -\frac{1}{V_0^p}\frac{\partial V^p}{\partial \sigma} \tag{3.15}$$

where superscript p refers to pores. Using similar notions from Sect. 3.2.1, we have

$$V_0^p = V_{0,e}^p + V_{0,t} \tag{3.16}$$

$$V^p = V_e^P + V_t \tag{3.17}$$

In the above two equations, V_t is considered to be a portion of pore volume (e.g., micro-cracks) in a rock body. Following the same procedure used to derive Eqs. 3.4 and 3.5, we obtain

$$dV_e^p = -C_e V_{0,e}^p d\sigma \tag{3.18}$$

$$dV_t = -\frac{V_{0,t}}{K_t}\exp\left(-\frac{\sigma}{K_t}\right)d\sigma \tag{3.19}$$

where C_e is the compressibility for the hard fraction of pore volume where engineering strain is applicable.

Combining Eqs. 3.15–3.19 yields (Liu et al. 2009)

$$C_{pc} = C_{pc}^{\infty} + \frac{\gamma_t}{\phi_0 K_t}\exp\left(-\frac{\sigma}{K_t}\right) \tag{3.20}$$

where

$$C_{pc}^{\infty} = C_e \frac{V_{0,e}^p}{V_0^p} \tag{3.21}$$

$$\phi_0 = \frac{V_0^p}{V_0} \tag{3.22}$$

Figure 3.3 shows the match of Eq. 3.20 to a compressibility data set presented in Jaeger et al. (2007). The data set was derived from a measured relationship between pore strain and confining stress for a Frio sandstone from East Texas [after

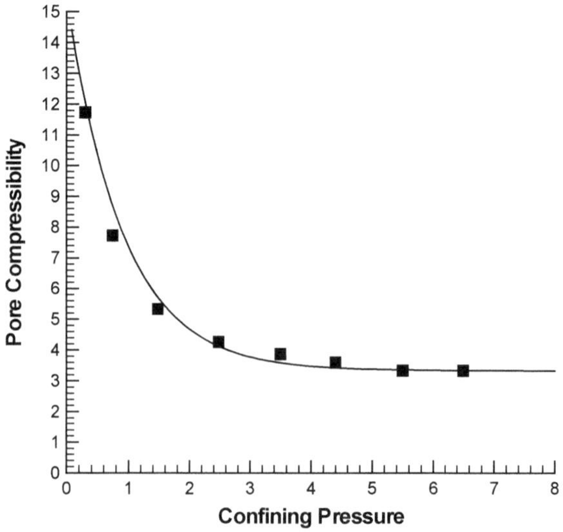

Fig. 3.3 Match of Eq. 3.20 to the data points of pore compressibility (10^{-6}/psi) as a function of confining pressure (or stress) (thousand psi) presented in Jaeger et al. (2007) (Liu et al. 2009) (reproduced by permission of Elsevier)

Carpenter and Spencer (1940)]. The satisfactory match indicates that our derived result is able to capture the key features of these experimental observations. The fitted parameter values are: $C_{pc}^{\infty} = 3.33 \times 10^{-6}$ psi^{-1} = 4.83×10^{-4} MPa^{-1}, $K_t = 1.1 \times 10^3$ psi = 7.6 MPa, and $\frac{\gamma_t}{\phi_0} = 0.011$. For typical porosity (ϕ_0) values of 10–20 % for a sandstone, the γ_t value from pore compressibility data ranges from 0.11 to 0.22 %, which again supports the notion that the so-called "soft" part is only a small percentage of pore volume. These parameter values are reasonably close to those obtained from rock compressibility data provided in the previous subsection.

3.2.2.3 Rock Porosity

Accurate descriptions of stress-dependent behavior for rock porosity (ϕ) are especially important for modeling coupled hydrological and mechanical processes, because the porosity is a critical parameter for fluid flow in porous rock. In this subsection, we derive a theoretical relationship between rock porosity and stress based on the TPHM.

Using the same notations as in the previous subsections and by definition of rock porosity, we have

$$d\phi = \frac{dV^p}{V} = \frac{dV_e^p + dV_t}{V} \tag{3.23}$$

where V is the bulk volume of porous rock. Note that the above equation ignores the effect of V change with stress on porosity change, considering that for most practical applications, V can be approximated by the unstressed volume V_0 for calculating rock porosity. Using this approximation and Eqs. 3.18 and 3.19, we obtain

$$d\phi = -\phi_e C_e d\sigma - \frac{\gamma_t}{K_t}\exp\left(-\frac{\sigma}{K_t}\right)d\sigma \tag{3.24}$$

where

$$\phi_e = \phi_0 - \gamma_t \tag{3.25}$$

Integrating Eq. 3.24 and using $\phi = \phi_0$ for $\sigma = 0$ gives (Liu et al. 2009)

$$\phi = \phi_e(1 - C_e\sigma) + \gamma_t \exp\left(\frac{\sigma}{K_t}\right) \tag{3.26}$$

The first term on the right hand side of Eq. 3.26 results from the hard part and the second term from the soft part. When the term $C_e\sigma$ is much smaller than one, the above equation can be approximately reduced to

$$\phi = \phi_e + \gamma_t \exp\left(-\frac{\sigma}{K_t}\right) \tag{3.27}$$

Rutqvist et al. (2002), based on laboratory experiments on sandstone by Davis and Davis (1999), proposed an empirical stress-porosity expression that is identical in form to Eq. 3.27. Similar empirical expressions were originally reported by Athy (1930) and further discussed in Neuzil (2003). Equation 3.26 is a more general stress-porosity relation and will be evaluated using data sets of Coyner (1984) who reported measured porosity-confining pressure relations for several types of rocks. His laboratory measurements for Berea sandstone and Weber sandstone samples are employed here, because these rock samples exhibit a relatively large degree of stress dependence of porosity. Note that confining pressure is used here to approximate the effective stress based on the test conditions (Berryman 1992; Liu et al. 2009).

Equation 3.26 includes four parameters: ϕ_e, C_e, γ_t and K_t. To avoid the non-uniqueness of parameter estimation from curve fitting, Liu et al. (2009) employed a simple procedure to determine parameter values from porosity versus confining pressure data. As shown in Figs. 3.4 and 3.5, the measured porosity is a well-defined linear function of the confining pressure for relatively high pressures (stresses). The slope of the straight line is used to determine $\phi_e C_e$ because the second term on the right hand side of Eq. 3.26, as a result of micro-crack closing, is negligible for high stress values. The porosity value at the intersection between the straight line and the vertical axis in Figs. 3.4 or 3.5 gives ϕ_e value corresponding to the value for the first term (at zero effective stress or confining pressure in this case) on the right hand side of Eq. 3.26. The measured porosity value at the zero confining pressure is equal to $\phi_e + \gamma_t$, as suggested from Eq. 3.26. The above

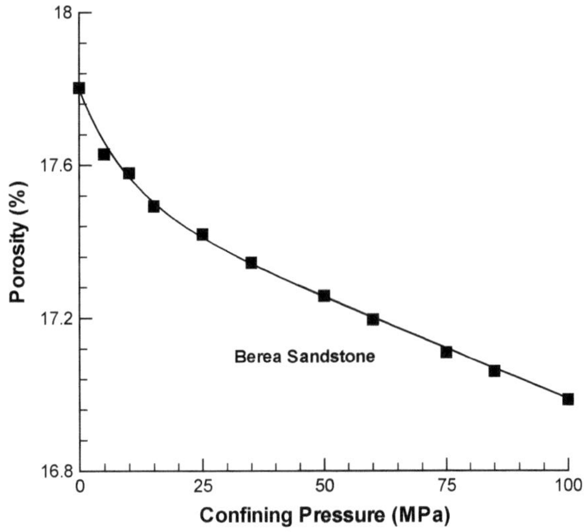

Fig. 3.4 Match between porosity values (as a function of confining pressure) calculated from Eq. 3.26 and the measured data for Berea Sandstone from Coyner (1984) (Liu et al. 2009) (reproduced by permission of Elsevier)

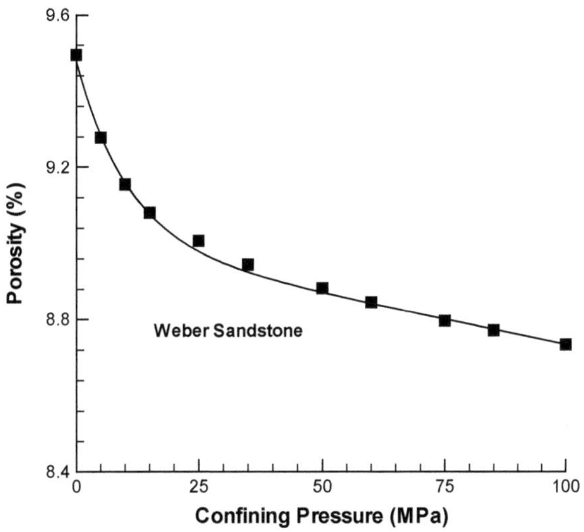

Fig. 3.5 Match between porosity values (as a function of confining pressure) calculated from Eq. 3.26 and the measured data for Weber Sandstone from Coyner (1984) (Liu et al. 2009) (reproduced by permission of Elsevier)

Table 3.1 Fitted parameter values from the experiment data sets of Coyner (1984)

Rock type	Parameter			
	ϕ_e (%)	C_e (MPa^{-1})	γ_t (%)	K_t (MPa)
Berea sandstone	17.52	3.04×10^{-4}	0.28	9.97
Weber sandstone	9.00	2.96×10^{-4}	0.48	10.60

Parameter values are from Liu et al. (2009)

procedure allows for direct determination of values for ϕ_e, C_e, and γ_t. The value for parameter K_t can be estimated using porosity data at relatively low confining pressures corresponding to the non-linear regime. Here, the K_t value is simply calculated from Eq. 3.26 using measured porosity value at a pressure of 10 MPa.

Satisfactory matches between results calculated from Eq. 3.26 (with the determined parameter values) and porosity data are shown in Figs. 3.4 and 3.5. The estimated parameter values for Berea and Weber sandstones are presented in Table 3.1. The estimated values for γ_t and K_t are generally consistent with those reported in the previous subsections where they are estimated from different types of data, suggesting that these two parameters (introduced in the TPHM) are very well defined and experimentally robust.

To further validate the TPHM, Liu et al. (2009) used parameter values estimated from the porosity data to calculate relations between the bulk-modulus and the pressure (stress), with a focus on checking if the calculated results could match the measured bulk modulus data for the same rock samples used for measuring porosity values (Coyner 1984). The stress-strain relationship (Eq. 3.8) gives the bulk modulus K as

$$K = \frac{1}{\frac{\gamma_e}{K_e} + \frac{\gamma_t}{K_t}\exp\left(-\frac{\sigma}{K_t}\right)} \quad (3.28)$$

The above equation can also be derived from Eq. 3.12. Values for $\frac{K_e}{\gamma_e}$ correspond to K values for high stresses and must be determined from measured K data; they cannot be directly determined from the porosity data. Based on the data shown in Figs. 3.6 and 3.7, the value of $\frac{K_e}{\gamma_e}$ is set to 13.5 GPa for Berea sandstone and 17.5

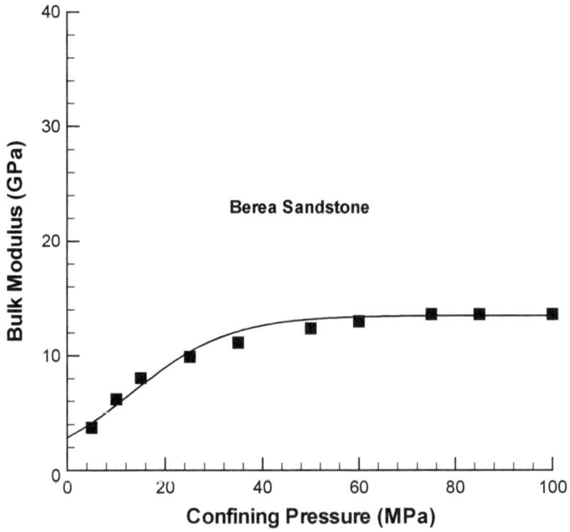

Fig. 3.6 A comparison between bulk modulus values (as a function of confining pressure) calculated from Eq. 3.26 and the measured data for Berea Sandstone from Coyner (1984) (Liu et al. 2009) (reproduced by permission of Elsevier)

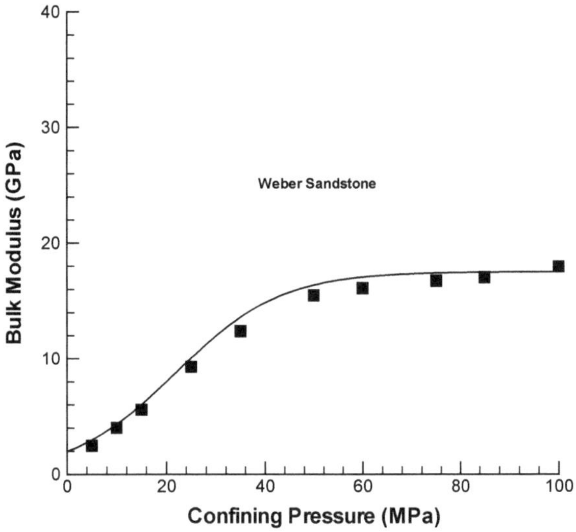

Fig. 3.7 A comparison between bulk modulus values (as a function of confining pressure) calculated from Eq. 3.26 and the measured data for Weber Sandstone from Coyner (1984) (Liu et al. 2009) (reproduced by permission of Elsevier)

GPa for Weber sandstone. These figures show that the calculated K curves are in a good agreement with the data. Considering that a curve-fitting procedure is not used there, the good agreement is encouraging and also further supports the robustness of the TPHM.

A formula to describe the stress sensitivity of the bulk modulus was also proposed by MacBeth (2004). He assumed (1) that the stress dependence behavior is a result of small-scale flaws or elements of compliance (Sayers and Kachanov 1995; Schoenberg and Sayers 1995) and (2) that the stress dependence of compliance is empirically approximated by an exponential function of stress. His formula for the bulk modulus is essentially the same as Eq. 3.28, and was successfully used to fit 179 sets of laboratory measurements on reservoir cores and outcrop sandstones that have low to moderate porosity and a range of clay fractions and cementation. This again supports the validity of Eq. 3.28 derived from the TPHM.

3.2.2.4 Relationship Between Permeability and Porosity for Low-Permeability Rock

In this subsection, we discuss some currently available relationships between permeability (k) and porosity (ϕ) and then present the TPHM-based relationship. The content of this subsection is based on the materials from Zheng et al. (2015).

One commonly used relationship between permeability and porosity of a rock has been empirically expressed by a power law (e.g., David et al. 1994; Ghabezloo et al. 2009b; Dong et al. 2010):

$$k/k_0 = (\phi/\phi_0)^m \tag{3.29}$$

where k_0 and ϕ_0 are permeability and porosity under the zero stress condition, respectively, and m is a material constant. David et al. (1994), using Eq. 3.29, analyzed permeability and porosity data under different effective stresses for five types of sandstones and also reviewed several published data sets on permeability-porosity relations for different geomaterials. They found that values for the exponent m range from 1.11 to 25.4. Ghabezloo et al. (2009b) measured permeability-porosity relation for an oil-well cement paste and showed that the observed relation obeys a power law identical to Eq. 3.29 with an m value equal to 11.00. Dong et al. (2010) conducted a series of permeability and porosity tests under different effective stresses and reported that their data are consistent with Eq. 3.29 with exponent m values ranging from 2.04 to 5.04 for fine-grained sandstone and from 9.92 to 70.17 for silty-shale samples.

The observed large values of the exponent m for most low-permeability rock mean that a small reduction in porosity (due to the applied stress) causes a significant decrease in permeability. However, there is not any physical justification for such large m values up to 70.17 for a porous medium characterized by a single continuum. The use of the total porosity in Eq. 3.29 essentially assumes that all pores, including micro-cracks and the alike, equally contribute to permeability, although micro-cracks are actually the decisive factor of pore connectivity for a low-permeability rock, as indicated by a number of researchers. For example, Byrnes (1997) and Byrnes and Castle (2000) found that average pore-throat sizes in some low-permeability sandstones under in situ stress decrease by as much as 50–70 % from the zero-stress condition, while the porosities under in situ stress are just several percent less. The micro-cracks and crack-like interconnecting throats generally constitute only a small portion of porosity, but can make a significant contribution to permeability (Byrnes 1997; Byrnes and Castle 2000). Since the relatively compliant part of rock body (e.g., micro-cracks) acts as critical fluid-flow paths, more attention should be paid to its deformation in determining hydraulic properties. However, almost all of the previous studies (Gangi 1978; Shi and Wang 1986; Dewhurst et al. 1998; Ghabezloo et al. 2009a; Dong et al. 2010) ignored the fact that different parts of rock have different mechanical responses under stress and at the same time affect the bulk permeability differently.

Micro-cracks are generally more deformable and thus experience a relatively large degree of deformation in a low effective stress range. Consequently, the impropriety of homogenizing the contributions of different types of pores to permeability (e.g. Eq. 3.29) becomes more pronounced for a low-permeability rock under a low effective stress. [For a high-permeability rock, this may not be a significant issue because the contribution of micro-cracks to the bulk permeability is relatively small anyway, as demonstrated in Dong et al. (2010).] It is therefore necessary to separate contributions of micro-cracks (and the alike) and the other pores for a low-permeability rock. The division of the hard and soft parts in the THPM provides a framework to consider such heterogeneity in developing stress-dependent permeability and porosity relationships. The TPHM-based derivations of these relationships are given below.

In the TPHM, micro-cracks and the alike correspond to the soft part, while the rest of pores belongs to the hard part. As indicated in Eq. 3.26, the porosity of a rock sample is the summation of the hard part porosity, i.e., $\phi_e = \phi_{e,0}(1 - C_e\sigma)$ that has a linear relationship with the effective stress, and the soft part porosity, i.e., $\phi_t = \gamma_t \exp\left(-\frac{\sigma}{K_t}\right)$ that changes exponentially with effective stress. The two porosities experience different degrees of reduction with increasing effective stress. Specifically, the soft part is more important in the low-stress range, but negligible in the high stress range because K_t is relatively small such that $\exp\left(-\frac{\sigma}{K_t}\right) \approx 0$ for $\sigma \gg K_t$.

The division of total porosity into the soft part and the hard part (that obey different relationships with effective stress within the TPHM framework) allows for consideration of their respective effects on the bulk permeability. Permeability changes in the relatively high effective-stress range are mainly controlled by the hard part, considering that the soft part porosity could be neglected in this stress range owing to the micro-crack closure. Mathematically, the stress-dependent permeability contributed by the hard part (referred to as "hard part permeability" hereafter) could be empirically given as:

$$k_e = k_{e,0}\exp\left[\beta\left(\phi_e - \phi_{e,0}\right)\right] = k_{e,0}\exp\left[-\beta C_e \phi_{e,0}\sigma\right] \tag{3.30}$$

where ϕ_e and k_e are the stress-dependent hard-part porosity and permeability, respectively, and β is a constant that represents a stress sensitive coefficient. Equation 3.30 seems to be consistent with most of the experimental observations that generally show a linear relationship between the logarithm of permeability and the effective stress in the relatively high effective stress range (David et al. 1994, 2001; Evans et al. 1997; Kwon et al. 2012).

The stress-sensitive permeability changes in the low effective stress range are mainly caused by the deformation of the soft part, although the soft part porosity makes up only a small portion of the total pore volume. The permeability contributed by the soft part, referred to as the "soft part permeability" hereafter, could be approximately considered as the bulk (total) permeability minus the hard part permeability:

$$k_t = k - k_e \tag{3.31}$$

where k is the total permeability and k_t is the soft part permeability. Mathematically, the relationship between soft part porosity and soft part permeability may be approximated as:

$$k_t = \alpha\phi_t^m \tag{3.32}$$

where ϕ_t and k_t are the soft part porosity and permeability, respectively, and α and m are material constants. Note that parameter m here is different from that in Eq. 3.29.

Combining Eqs. 3.26, 3.31 and 3.32 yields the total permeability as (Zheng et al. 2015):

$$k = k_{e,0}\exp\left[-\beta C_e \phi_{e,0}\sigma\right] + \alpha\left[\gamma_t \exp\left(-\frac{\sigma}{K_t}\right)\right]^m \tag{3.33}$$

As previously mentioned, the soft part corresponds to those thin and slit-like micro-cracks and the alike. If such a hypothesis holds, the permeability change caused by the soft part deformation should obey the "cubic law" that defines the relationship between bulk fracture permeability and the corresponding average fracture aperture (Witherspoon et al. 1980; Zimmerman and Bodvarsson 1996; Kwon et al. 2012). In other words, the soft part permeability should be well related to the soft part porosity through a relationship similar to the "cubic law", i.e., m in Eq. 3.32, depending on how micro-cracks are connected with each other and with pores, should not be very far from 3. Note that the average fracture aperture corresponds to the soft-part porosity. Zheng et al. (2015) evaluated the proposed relationships using the experimental observations collected from the literature.

A number of researchers have reported stress-dependent permeability data for low-permeability rock in the literature (e.g., McLatchie et al. 1958; Vairogs et al. 1971; Thomas and Ward 1972; Walls et al. 1982; Kilmer et al. 1987; Brighenti 1989; Spencer 1989; Kwon et al. 2012; Lei et al. 2007; Jasinge et al. 2011; Konecny and Kozusnikova 2011; Metwally and Sondergeld 2011). However, one also needs stress-dependent porosity data for evaluating the relationships discussed above. To the best of our knowledge, only a few researchers provided the data sets of the both types for a same rock (Wyble 1958; Jones and Owens 1980; Yale and Nur 1985; David et al. 1994, Mohiuddin et al. 2000; Dong et al. 2010). In the selection of the suitable experimental data sets for the evaluation, several criteria are considered by Zheng et al. (2015). Firstly, the samples and the associated measurement procedures need to be well documented such that enough information is available for checking the reliability of the data. Secondly, the working fluid to measure the permeability is gas rather than liquid, because gas (e.g., nitrogen and helium) is more chemically inert than liquid (e.g., pure water and NaCl solution). Thirdly, the samples collected from the subsurface are preferred over outcrop rock samples, since the relationships of interest are mostly for underground engineering applications. Other considerations include sufficient data density and microscopic pore structure analysis. Given these considerations, Zheng et al. (2015) selected the data set of Dong et al. (2010) to evaluate the proposed relationships.

In the study of Dong et al. (2010), rock samples were collected from the depths of 900–1235 m as a part of a deep drilling project (Taiwan Chelungpu fault Drilling Project, TCDP-A) in the Western Foothills of Taiwan. Relatively homogeneous cores were selected by Dong et al. (2010) for their study. They prepared for the samples carefully to reduce the occurrence of obvious cracks during sample preparation and conducted a series of stress-dependent porosity and permeability tests for the dry rock samples on an integrated porosity and permeability measurement system. A steady state flow method was employed to measure the sample permeability under different stresses while stress-dependent porosity was estimated by a gas expansion method. Nitrogen was used as the test fluid for both tests. In the experiments, the confining pressure were first gradually increased (loading) from 3 to 5 MPa, then to 20 MPa (in 5 MPa increments), and finally to 120 MPa (in 10 MPa increments). The confining pressure was then gradually reduced (unloaded) back to 3 MPa in the reverse order. The average pore pressures are relatively low compared with confining pressures; they are 0.13–1.40 MPa for the permeability measurement and 0.3–1.41 MPa for the porosity measurement. The effective stress is calculated as the difference between confining pressure and pore pressure. There are two types of tested samples, i.e., Pliocene to Pleistocene fine-grained sandstone and silty-shale. In this subsection, we focus on the stress-dependent hydraulic properties of the silty-shale that has relatively low permeability. The related geophysical properties of the silty-shale samples are listed in Table 3.2.

Table 3.2 Basic geophysical properties of the silty-shale samples (Zheng et al. 2015)

Sample name	Depth(m)	Dry density (g/cm^3)	Rock type
R255_sec2	902.68	2.59	Silty-shale
R287_sec1	972.42	2.58	Silty-shale
R351_sec2	1114.33	2.59	Silty-shale
R390_sec3	1174.24	2.66	Silty-shale

Reproduced by permission of Elsevier

Since the porosity under zero effective stress ($\phi_{e,0}$) was not available from Dong et al. (2010), we rewrite Eq. 3.26 as (Zheng et al. 2015):

$$\phi = \phi_{e,l}(1 - C_e\Delta\sigma) + \gamma_{t,l}\exp\left(-\frac{\Delta\sigma}{K_t}\right) \tag{3.34}$$

where $\phi_{e,l}$ is the hard-part porosity at the lowest observed effective stress σ_l and related to $\phi_{e,0}$ by $\phi_{e,l} = \phi_{e,0}(1 - C_e\sigma_l)$, the subscript l here refers to the lowest effective stress state, $\Delta\sigma = \sigma - \sigma_l$, and $\gamma_{t,l}$ is the soft-part volumetric fraction at σ_l and related to γ_t by $\gamma_t = \gamma_{t,l}\exp(\sigma_l/K_t)$. Note that Eq. 3.34 is an approximation relation because its derivation uses $\phi_{e,l} - \phi_{e,0}C_e\Delta\sigma \approx \phi_{e,l}(1 - C_e\Delta\sigma)$. This is a reasonable approximation when σ_l is close to zero.

Similarly, Eq. 3.33 can be rewritten as (Zheng et al. 2015):

$$k = k_{e,l}\exp\left[-\beta C_e\phi_{e,l}\Delta\sigma\right] + \alpha\left[\gamma_{t,l}\exp\left(-\frac{\Delta\sigma}{K_t}\right)\right]^m \tag{3.35}$$

where $k_{e,l}$ is the permeability at the stress σ_l and related to k_e by $k_e = k_{e,l}\exp\left(\beta C_e\phi_{e,l}\sigma_l\right)$.

The procedure to determine parameters in Eq. 3.34, by matching the stress-dependent porosity data (e.g., Fig. 3.8a), is already described in Sect. 3.2.2.3. Similar procedure is used to determine the related parameters in Eq. 3.35. For example, Fig. 3.8b shows a permeability-effective stress relation and a linear relation exists for $\log(K_e)$ versus σ for the relatively high effective stress range, as shown by the red straight line in the figure. In the linear regime, the contribution of the soft part to the bulk permeability can be neglected, because the second term on the right hand side of Eq. 3.35 approaches zero. The slope of the straight line gives the value of $\left(-\beta C_e\phi_{e,l}\right)$ that is used for the determination of β. Note that $\phi_{e,l}$ and C_e are determined by fitting the porosity-stress data with Eq. 3.34. Parameter $k_{e,l}$ can then be obtained as the permeability on the straight line at $\sigma = \sigma_l$. After the hard part permeability is known, the soft part permeability is approximated by the total or bulk permeability minus the hard part permeability. The use of Eq. 3.32 to fit the estimated relation between the soft part permeability and the soft part porosity

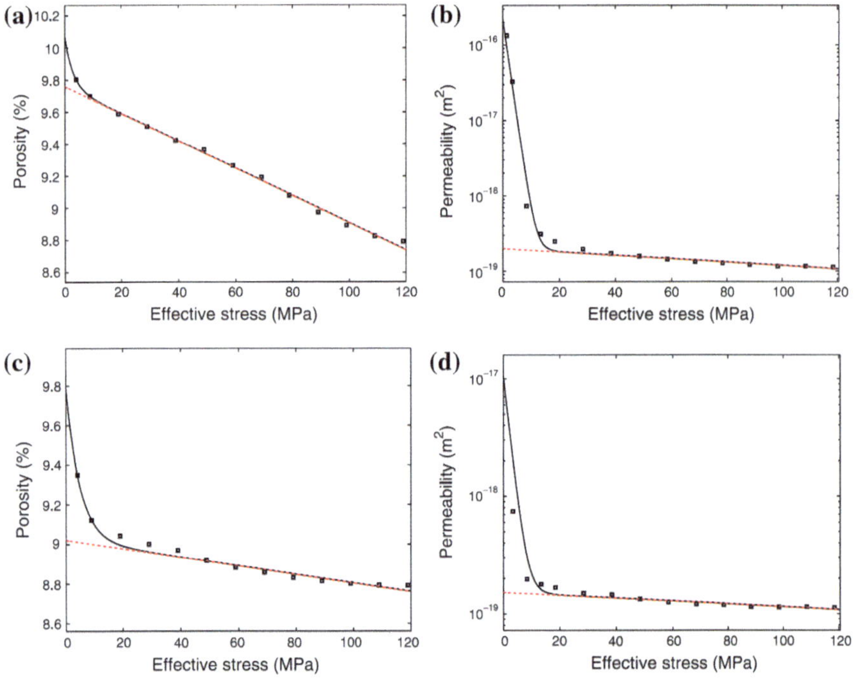

Fig. 3.8 The matching result of the proposed relationships and the test data for sample R255_sec2 (Zheng et al. 2015): **a** porosity-stress (loading stage), **b** permeability-stress permeability (loading stage), **c** porosity-stress (unloading stage), and **d** permeability-stress (unloading stage) (reproduced by permission of Elsevier)

yields the corresponding value for m. The m values for different samples are given in Figs. 3.8, 3.9, 3.10 and 3.11.

Reasonable matches between Eqs. 3.34 and 3.35 and the experimental data, shown in Figs. 3.8, 3.9, 3.10 and 3.11, support the validity of these TPHM-based porosity-stress and permeability-stress relationships. As shown in the figures for the porosity-stress relation, the soft part mainly accounts for the nonlinear porosity reduction with effective stress in the low stress range and the corresponding porosity reduction in this range is largely a result of the closing process of slit-like micro-cracks. In addition, as shown in the figures for the permeability-stress relation, changes in the soft part permeability significantly contribute to the overall permeability reduction in the low effective stress range. While the soft part porosity is only a small portion of the total pore volume, the soft part acts as critical flow paths for fluid flow in a low-permeability medium and therefore its deformation significantly affects the permeability.

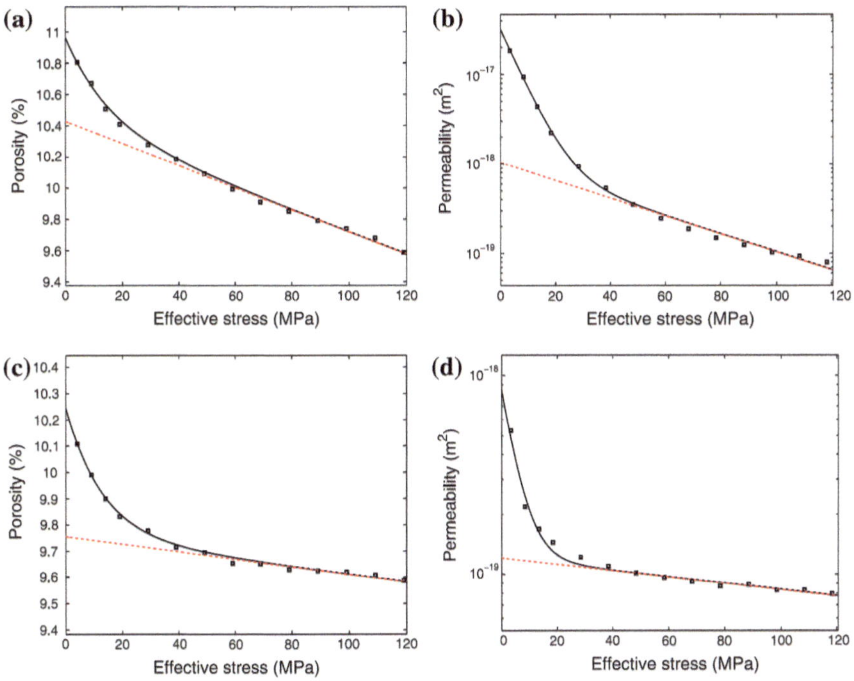

Fig. 3.9 The matching result of the proposed relationships and the test data for sample R287_sec1 (Zheng et al. 2015): **a** stress-porosity (loading stage), **b** stress-permeability (loading stage), **c** stress-porosity (unloading stage), and **d** stress-permeability (unloading stage) (reproduced by permission of Elsevier)

The determined values of $\phi_{e,l}$, C_e, $\gamma_{t,l}$, K_t, $k_{e,l}$, β, α and m are listed in Table 3.3. As expected, the soft part porosity at $\sigma = \sigma_l$, $\gamma_{t,l}$, is generally small (0.07–0.77 %). The elastic modulus of the soft part ranges from 2.83 to 14.81 MPa that are significantly smaller than that for the hard part. The values for $k_{e,l}$ are much smaller than the total permeability at $\sigma = \sigma_l$. In addition, the observed relationships between the soft part permeability and the soft part porosity are reasonably fitted by a power law relation with average exponent m values of 2.28 for loading process and 2.04 for unloading process. These m values are not very far from 3 (corresponding to the cubic law) and therefore generally support the hypothesis that the soft part permeability is controlled by the slit-like micro-cracks and the alike. The discrepancy between the estimated m values and 3 is very likely a result of the fact that effects of interactions between the hard and soft parts, for simplicity, are ignored here when calculating permeability for the two parts.

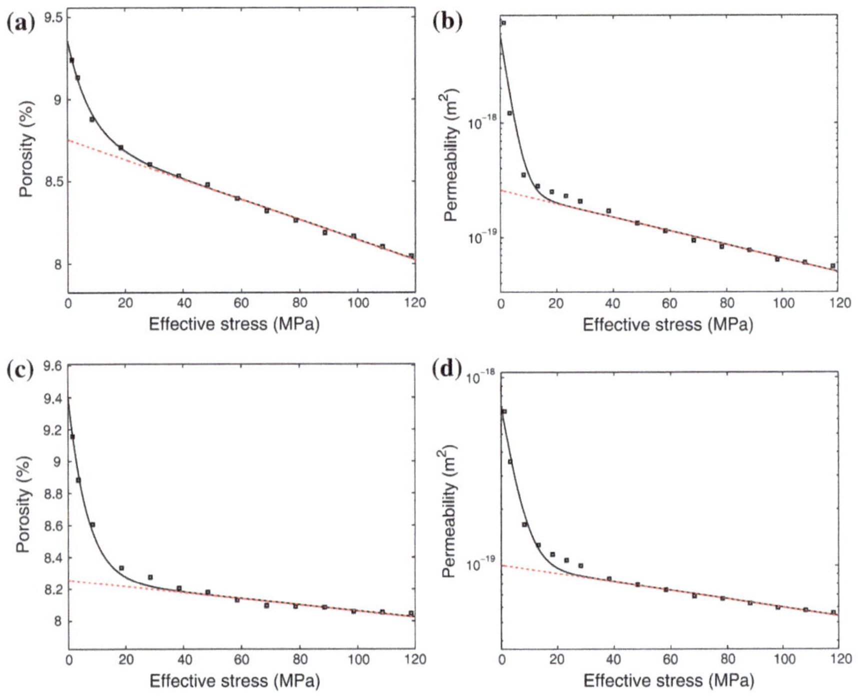

Fig. 3.10 The matching result of the proposed relationships and the test data for sample R351_sec2 (Zheng et al. 2015): **a** stress-porosity (loading stage), **b** stress-permeability (loading stage), **c** stress-porosity (unloading stage), and **d** stress-permeability (unloading stage) (Reproduced by permission of Elsevier)

3.2.3 TPHM for Anisotropic Stress Condition

In Sect. 3.2.2, we discussed the TPHM for the isotropic stress condition. This section is devoted to the TPHM for the anisotropic stress condition. Note that in the both sections, the mechanical properties are isotropic. Without losing generality, we consider stress-strain relationships corresponding to the three principal stresses (Fig. 3.12). The content of this subsection is based on the materials from Zhao and Liu (2012).

To extend the TPHM to the anisotropic stress condition, Zhao and Liu (2012) assumed, for the convenience of mathematical development, that the principal strain resulting from the soft part is a function of the principal stress along the same direction only and has nothing to do with the other principal stresses. The validity of this assumption will be evaluated by comparing the related theoretic results with experimental observations.

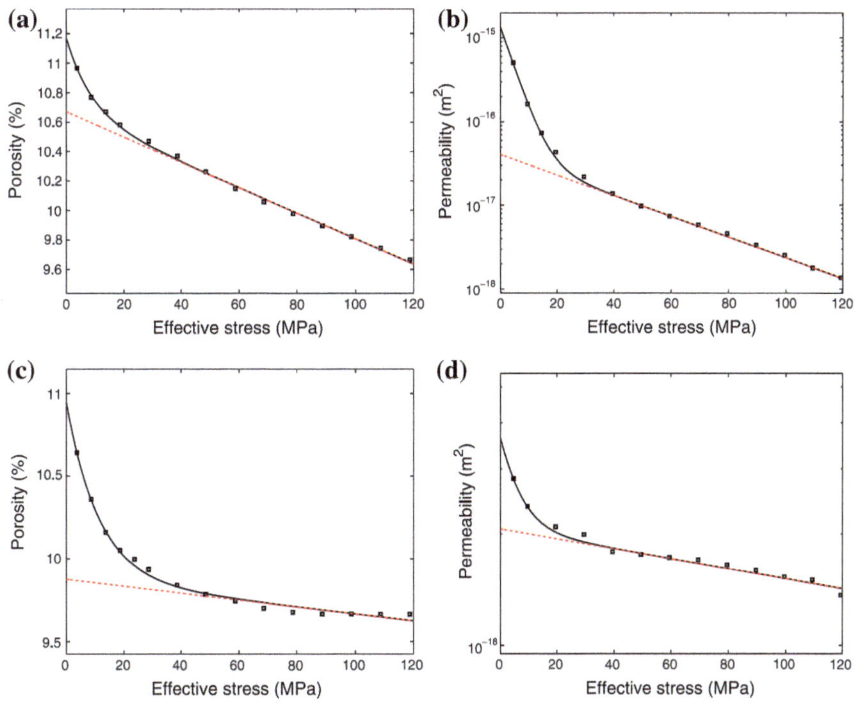

Fig. 3.11 The matching result of the proposed relationships and the test data for sample R351_sec3 (Zheng et al. 2015): **a** stress-porosity (loading stage), **b** stress-permeability (loading stage), **c** stress-porosity (unloading stage), and **d** stress-permeability (unloading stage) (reproduced by permission of Elsevier)

Following the procedure to derive Eq. 3.8, Zhao and Liu (2012) derive the following expressions for the principal strains:

$$
\begin{aligned}
d\varepsilon_1 &= \frac{\gamma_e'}{E_e}[d\sigma_1 - v(d\sigma_2 + d\sigma_3)] + \frac{\gamma_t'}{E_t}\exp\left(-\frac{\sigma_1}{E_t}\right)d\sigma_1 \\
d\varepsilon_2 &= \frac{\gamma_e'}{E_e}[d\sigma_2 - v(d\sigma_1 + d\sigma_3)] + \frac{\gamma_t'}{E_t}\exp\left(-\frac{\sigma_2}{E_t}\right)d\sigma_2 \\
d\varepsilon_3 &= \frac{\gamma_e'}{E_e}[d\sigma_3 - v(d\sigma_1 + d\sigma_2)] + \frac{\gamma_t'}{E_t}\exp\left(-\frac{\sigma_3}{E_t}\right)d\sigma_3
\end{aligned} \tag{3.36}
$$

$$
d\varepsilon_i = -\frac{dl^i}{l_0^i} \quad (\mathrm{i} = 1, 2, 3) \tag{3.37}
$$

$$
dl^i = dl_e^i + dl_t^i \quad (\mathrm{i} = 1, 2, 3) \tag{3.38}
$$

Table 3.3 Fitted parameter values from the experimental data of Dong et al. (2010) (Zheng et al. 2015)

Sample		$\phi_{e,l}$ (%)	$\gamma_{t,l}$ (%)	C_e (MPa^{-1})	K_t (MPa)	$k_{e,l}$ (m^2)	β	α (m^2)	m
R255_sec2	Loading	9.73	0.07	8.75×10^{-4}	2.83	1.97×10^{-19}	0.61	1.66×10^{-15}	1.70
	Unloading	9.01	0.34	2.41×10^{-4}	4.84	1.51×10^{-19}	1.30	1.99×10^{-17}	2.41
R287_sec1	Loading	10.40	0.41	6.81×10^{-4}	14.81	9.54×10^{-19}	3.25	1.32×10^{-16}	2.34
	Unloading	9.75	0.36	1.48×10^{-4}	13.04	1.19×10^{-19}	2.53	5.53×10^{-18}	2.65
R351_sec2	Loading	8.75	0.49	6.97×10^{-4}	8.09	2.54×10^{-19}	2.24	2.61×10^{-17}	3.03
	Unloading	8.26	0.90	2.35×10^{-4}	6.67	1.01×10^{-19}	2.68	4.12×10^{-19}	1.54
R390_sec3	Loading	10.64	0.33	8.11×10^{-4}	8.74	3.66×10^{-17}	3.30	5.42×10^{-15}	2.04
	Unloading	9.87	0.77	2.12×10^{-4}	11.30	2.04×10^{-18}	1.49	1.33×10^{-18}	1.62

Fig. 3.12 Principal stresses

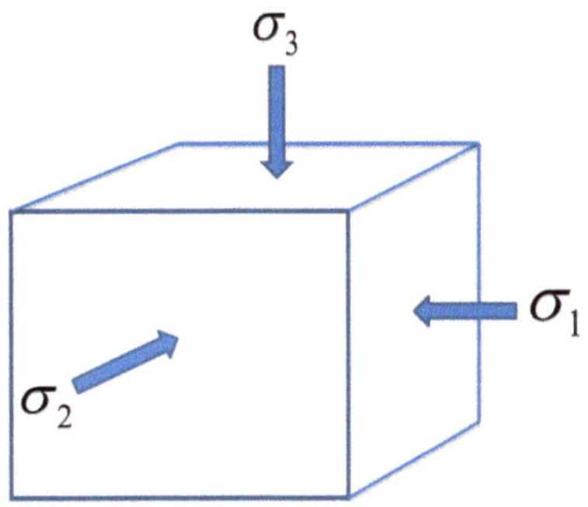

$$l_0^i = l_{0,e}^i + l_{0,t}^i \quad (\mathrm{i} = 1, 2, 3) \tag{3.39}$$

$$\gamma_t' = \frac{l_{0,t}^i}{l_0^i} \quad (\mathrm{i} = 1, 2, 3) \tag{3.40}$$

$$\gamma_e' = 1 - \gamma_t' \quad (\mathrm{i} = 1, 2, 3) \tag{3.41}$$

where σ_1, σ_2, σ_3 are principal stresses, ε_1, ε_2, ε_3 are principal engineering strains, v is Poisson ratio for the hard part, γ_t' is the ratio of the soft-part length to the length for the entire rock body in one principal direction (under unstressed conditions), l^i is the rock length in i principal direction, and E_e and E_t refer to Young's (elastic) modulus for the hard and soft part, respectively. Again, we use subscripts 0, e, and t to denote the unstressed state, the hard part and the soft part, respectively, for a rock body.

The first term on the right hand side of Eq. 3.36 results from the hard part and the second term from the soft part. Without the second part, our developed strain-stress relationship is consistent with the general engineering-strain-based Hooke's Law (Jaeger et al. 2007).

To derive the relationship between the γ_t (Eq. 3.9) and γ_t' for the soft part, we consider a rock element whose volume V_0, by definition, is related to the length l_0^i (i = 1, 2, 3) by

$$V_0 = \left(l_{0,e}^1 + l_{0,t}^1\right)\left(l_{0,e}^2 + l_{0,t}^2\right)\left(l_{0,e}^3 + l_{0,t}^3\right) = l_{0,e}^1 l_{0,e}^2 l_{0,e}^3 + 3 l_0^1 l_0^2 l_0^3 \gamma_t' (1 - \gamma_t')^2 + 3 l_0^1 l_0^2 l_0^3 (\gamma_t')^2 (1 - \gamma_t') + l_0^1 l_0^2 l_0^3 (\gamma_t')^3 \tag{3.42-1}$$

Because γ_t' is generally much smaller than one, we can neglect the high-order terms of γ_t' in Eq. 3.42-1. Then V_0 can be written as:

$$V_0 \approx l_{0,e}^1 l_{0,e}^2 l_{0,e}^3 + 3\gamma_t' l_0^1 l_0^2 l_0^3 = V_{0,e} + 3\gamma_t' V_0 \tag{3.42-2}$$

Combining Eqs. 3.8, 3.9, 3.10, 3.42-1, and 3.42-2 yields (Zhao and Liu 2012):

$$\gamma_t' = \frac{\gamma_t}{3} \tag{3.43}$$

It is well known that the engineering volumetric strain is the sum of three principal strains (Jaeger et al. 2007):

$$d\varepsilon_v = -\frac{dV}{V_0} = d\varepsilon_1 + d\varepsilon_2 + d\varepsilon_3 \tag{3.44}$$

where ε_v is the volumetric strain. Combining Eqs. 3.36, 3.43 and 3.44 yields

$$\begin{aligned} d\varepsilon_v = & \frac{(3-\gamma_t)(1-2v)}{3E_e}(d\sigma_1 + d\sigma_2 + d\sigma_3) + \frac{\gamma_t}{3E_t}\exp\left(-\frac{\sigma_1}{E_t}\right)d\sigma_1 \\ & + \frac{\gamma_t}{3E_t}\exp\left(-\frac{\sigma_2}{E_t}\right)d\sigma_2 + \frac{\gamma_t}{3E_t}\exp\left(-\frac{\sigma_3}{E_t}\right)d\sigma_3 \end{aligned} \tag{3.45}$$

To make Eqs. 3.8 and 3.45 consistent under the isotropic stress condition ($\sigma_1 = \sigma_2 = \sigma_3$), we can relate the Young's modulus E_e to bulk modulus K_e by

$$K_e = \frac{E_e \gamma_e}{3(1-2v)\gamma_e'} = \frac{E_e(1-\gamma_t)}{3(1-2v)(1-\gamma_t')} \tag{3.46}$$

where γ_t and γ_t' are generally on the order of 10^{-2}, much smaller than one for most porous rocks and therefore can be neglected in Eq. 3.46. In this case, we obtain

$$K_e = \frac{E_e}{3(1-2v)} \tag{3.47}$$

Furthermore, consistency between Eqs. 3.8 and 3.45 requires

$$K_t = E_t \tag{3.48}$$

With the definitions given in Eqs. 3.46 and 3.47, Eq. 3.8 becomes a special case of Eq. 3.45.

Based on Eq. 3.43 and the condition that strains are zero under unstressed state, principal strains can be solved from Eq. 3.36 as (Zhao and Liu 2012)

$$\begin{aligned}
\varepsilon_1 &= \frac{(3-\gamma_t)}{3E_e}[\sigma_1 - \nu(\sigma_2+\sigma_3)] + \frac{\gamma_t}{3}\left[1-\exp\left(-\frac{\sigma_1}{E_t}\right)\right] \\
\varepsilon_2 &= \frac{(3-\gamma_t)}{3E_e}[\sigma_2 - \nu(\sigma_1+\sigma_3)] + \frac{\gamma_t}{3}\left[1-\exp\left(-\frac{\sigma_2}{E_t}\right)\right] \\
\varepsilon_3 &= \frac{(3-\gamma_t)}{3E_e}[\sigma_3 - \nu(\sigma_2+\sigma_1)] + \frac{\gamma_t}{3}\left[1-\exp\left(-\frac{\sigma_3}{E_t}\right)\right]
\end{aligned} \tag{3.49}$$

Note that the second term on the right hand side is from the soft part in Eqs. 3.36 and 3.49. That term, unlike the first term resulting from the hard part, is only a function of the corresponding principal stress and not related to other stresses. In other words, Poisson's ratio is ignored for the soft part in Eqs. 3.36 and 3.49.

The treatment of Poisson's ratio for the soft part, as the first step, is considered a rough approximation, and further research may be needed to refine the treatment (by incorporating Poisson's ratio for the soft part). Poisson's ratio is defined as the negative ratio of transverse strain to the longitudinal strain, under the uniaxial stress condition (Jaeger et al. 2007). Although the use of approximate or typical values in most rock-mechanics applications does not create significant problems, Poisson's ratio plays an undeniably important role in the elastic deformation of rocks and rock masses subjected to static or dynamic stresses. Furthermore, its effects emerge in a wide variety of rock engineering applications, ranging from basic laboratory tests on intact rocks to field measurements for in situ stresses or deformability of rock masses (Gercek 2007). Poisson (1829) recommended the value of Poisson's ratio as 1/4. To make Young's shear and bulk moduli of a material positive, the theoretical value of Poisson's ratio must lie in the range between −1 and ½ (Jaeger et al. 2007). According to Gercek (2007), the values of Poisson's ratio for many materials are between 0 and 0.5. For the case of rock, many factors, including the porosity and the geometry (size and the orientation), distribution, and connectivity of pores, are expected to influence the value of Poisson's ratio (Gercek 2007). Zhao and Liu (2012) assume that only the hard part has Poisson's effect. As demonstrated later in this section, this approximation may be adequate for most practical applications in rock mechanics.

Equations developed above are for the principal stress/strain coordinate system. The relationships between stress and strain in a general coordinate system (*x*, *y*, *z*) can be obtained through the coordinate transformation. Transformation of Eq. 3.49 from a principal stress/strain coordinate system to a general coordinate system, as shown in Fig. 3.13, yields (Poulos and Davis 1974):

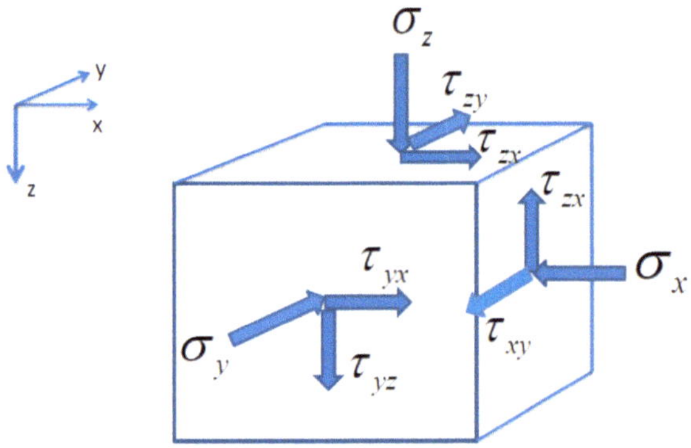

Fig. 3.13 Stress components in the general coordinate system. The components are shown only on the three sides of the rock element here

$$\begin{aligned}
\varepsilon_x &= \frac{(3-\gamma_t)}{3E_e}\left[\sigma_x - v\left(\sigma_y+\sigma_z\right)\right] + \frac{\gamma_t}{3}\left[1-\exp\left(-\frac{\sigma_1}{E_t}\right)l_1^2 - \exp\left(-\frac{\sigma_2}{E_t}\right)l_2^2 - \exp\left(-\frac{\sigma_3}{E_t}\right)l_3^2\right]\\
\varepsilon_y &= \frac{(3-\gamma_t)}{3E_e}\left[\sigma_y - v\left(\sigma_x+\sigma_z\right)\right] + \frac{\gamma_t}{3}\left[1-\exp\left(-\frac{\sigma_1}{E_t}\right)m_1^2 - \exp\left(-\frac{\sigma_2}{E_t}\right)m_2^2 - \exp\left(-\frac{\sigma_3}{E_t}\right)m_3^2\right]\\
\varepsilon_z &= \frac{(3-\gamma_t)}{3E_e}\left[\sigma_z - v\left(\sigma_x+\sigma_y\right)\right] + \frac{\gamma_t}{3}\left[1-\exp\left(-\frac{\sigma_1}{E_t}\right)n_1^2 - \exp\left(-\frac{\sigma_2}{E_t}\right)n_2^2 - \exp\left(-\frac{\sigma_3}{E_t}\right)n_3^2\right]
\end{aligned} \tag{3.50}$$

$$\begin{aligned}
\gamma_{xy} &= \frac{2(3-\gamma_t)(1+v)}{3E_e}\left[(\sigma_2-\sigma_1)l_2m_2 + (\sigma_3-\sigma_1)l_3m_3\right]\\
&\quad + \frac{2\gamma_t}{3}\left[\exp\left(-\frac{\sigma_1}{E_t}\right) - \exp\left(-\frac{\sigma_2}{E_t}\right)\right]l_2m_2 + \frac{2\gamma_t}{3}\left[\exp\left(-\frac{\sigma_1}{E_t}\right) - \exp\left(-\frac{\sigma_3}{E_t}\right)\right]l_3m_3\\
\gamma_{xz} &= \frac{2(3-\gamma_t)(1+v)}{3E_e}\left[(\sigma_2-\sigma_1)l_2n_2 + (\sigma_3-\sigma_1)l_3n_3\right]\\
&\quad + \frac{2\gamma_t}{3}\left[\exp\left(-\frac{\sigma_1}{E_t}\right) - \exp\left(-\frac{\sigma_2}{E_t}\right)\right]l_2n_2 + \frac{2\gamma_t}{3}\left[\exp\left(-\frac{\sigma_1}{E_t}\right) - \exp\left(-\frac{\sigma_3}{E_t}\right)\right]l_3n_3\\
\gamma_{yz} &= \frac{2(3-\gamma_t)(1+v)}{3E_e}\left[(\sigma_2-\sigma_1)m_2n_2 + (\sigma_3-\sigma_1)m_3n_3\right]\\
&\quad + \frac{2\gamma_t}{3}\left[\exp\left(-\frac{\sigma_1}{E_t}\right) - \exp\left(-\frac{\sigma_2}{E_t}\right)\right]m_2n_2 + \frac{2\gamma_t}{3}\left[\exp\left(-\frac{\sigma_1}{E_t}\right) - \exp\left(-\frac{\sigma_3}{E_t}\right)\right]m_3n_3
\end{aligned} \tag{3.51}$$

where $l_i = \cos(i, x)$, $m_i = \cos(i, y)$, $n_i = \cos(i, z)$ and i = 1, 2, 3 is the index for the direction of i principal stress. The functions cos(i, x), cos(i, y), and cos(i, z) are the cosine of the angles between i and x, y, and z directions, respectively, and are given as

$$\cos(i,x)=\frac{(\sigma_y-\sigma_i)(\sigma_z-\sigma_i)-\tau_{zy}\tau_{yz}}{\sqrt{\left[(\sigma_y-\sigma_i)(\sigma_z-\sigma_i)-\tau_{zy}\tau_{yz}\right]^2+\left[\tau_{zy}\tau_{xz}-\tau_{xy}(\sigma_z-\sigma_i)\right]^2+\left[\tau_{xy}\tau_{yz}-\tau_{xz}(\sigma_y-\sigma_i)\right]^2}}$$
$$\cos(i,y)=\frac{\tau_{zy}\tau_{xz}-\tau_{xy}(\sigma_z-\sigma_i)}{\sqrt{\left[(\sigma_y-\sigma_i)(\sigma_z-\sigma_i)-\tau_{zy}\tau_{yz}\right]^2+\left[\tau_{zy}\tau_{xz}-\tau_{xy}(\sigma_z-\sigma_i)\right]^2+\left[\tau_{xy}\tau_{yz}-\tau_{xz}(\sigma_y-\sigma_i)\right]^2}}$$
$$\cos(i,z)=\frac{\tau_{xy}\tau_{yz}-\tau_{xz}(\sigma_y-\sigma_i)}{\sqrt{\left[(\sigma_y-\sigma_i)(\sigma_z-\sigma_i)-\tau_{zy}\tau_{yz}\right]^2+\left[\tau_{zy}\tau_{xz}-\tau_{xy}(\sigma_z-\sigma_i)\right]^2+\left[\tau_{xy}\tau_{yz}-\tau_{xz}(\sigma_y-\sigma_i)\right]^2}} \tag{3.52}$$

To evaluate the validity of the developed stress-strain relationship (Eq. 3.49), Zhao and Liu (2012) used the relationship to fit the unconfined compression tests presented in Corkum and Martin (2007a) and Olalla et al. (1999) for Opalinus Clay rock. In Corkum and Martin (2007a), rock samples with a diameter of 83 mm were saw-cut from Boreholes BRA-1 and BRA-2, drilled using oil and air as drilling fluids, respectively (Corkum and Martin 2007a). In Olalla et al. (1999), rock samples were 78 mm in diameter.

To avoid (as much as possible) the non-uniqueness of parameter estimation from the curve fitting, Zhao and Liu (2012) used the procedure similar to that used in Sect. 3.2.2.3 (for analyzing stress-dependency of porosity) to estimate values for the parameters from stress-strain data. As shown in Fig. 3.14, measured relations between stress and strain are very well represented by a straight line for relatively high stresses. The slope of the straight line is used to determine $\frac{3E_e}{3-\gamma_t}=\frac{E_e}{\gamma'_e}$, because the exponential terms on the right hand side of Eq. 3.49 are negligible for high stresses. The strain value at the intersection between the straight line and the strain axis in Figs. 3.13 and 3.14 gives the $\gamma'_t=\frac{\gamma_t}{3}$ value, considering that the straight line represents the first term on the right-hand side of Eq. 3.49. The above procedure allows for direct determination of values for E_e, γ'_e and γ'_t. The remaining parameter E_t can be estimated using a data point at relatively low stress corresponding to the nonlinear regime.

As indicted in Fig. 3.14, the data are in excellent agreement with our theoretical results, suggesting that the assumption regarding Poisson's ratio for the soft part seems to be adequate. Fitted parameter values are given in Table 3.4. As an example, Fig. 3.15 also shows a comparison between hard-part and soft-part strain for specimen 9963. Several interesting observations can be made when comparing theoretical results with experimental observations. Firstly, the soft part has a larger strain than the hard part at an early stage of uniaxial loading, even though the volumetric ratio γ_t is much smaller than that for the hard part (Fig. 3.15). Secondly, when applied stress loading on the rock frame increases, the shape of pores in the soft-part changes, trending toward completely closure, while the hard-part pores remain hard and resist closure. Note that the Young's (elastic) modulus for the soft-part E_t ranges from 0.2 to 1.2 MPa, which is much smaller than the Young's (elastic) modulus for the hard-part E_e, which ranges from 2164.5 to 3345.1 MPa (Table 3.4). The difference between the Young's (elastic) moduli indicates that the

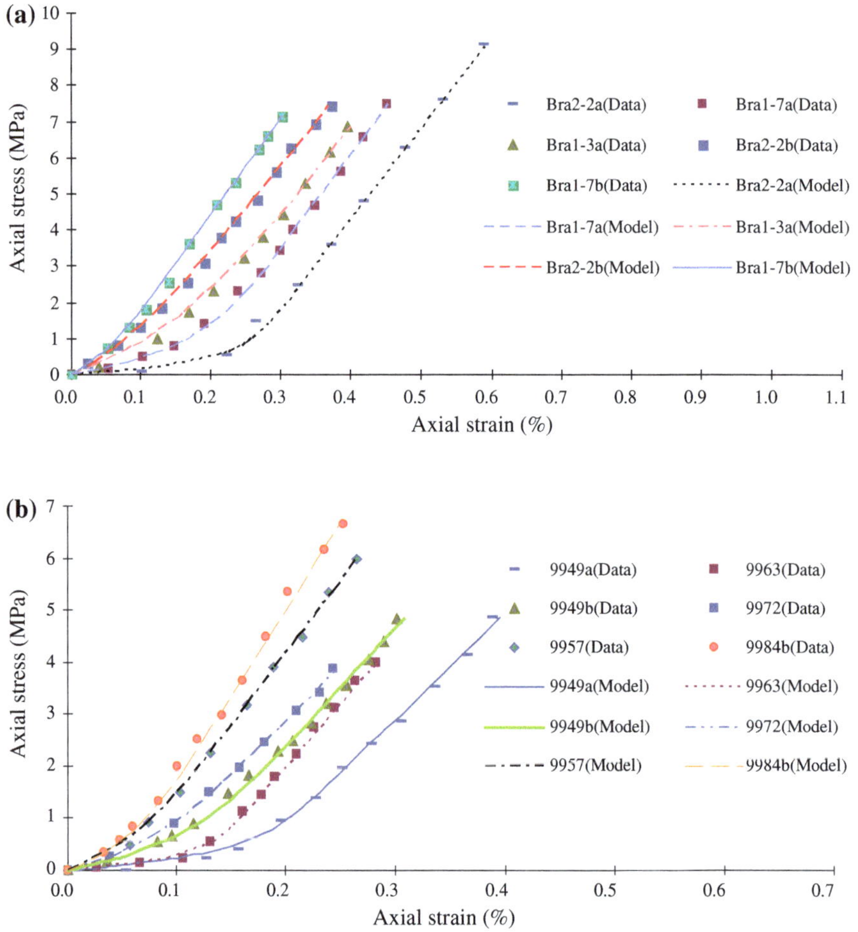

Fig. 3.14 Matches between results calculated from Eq. 3.49 and experimental data from unconfined compression tests on clay rock reported by **a** Corkum and Martin (2007a) and **b** Olalla et al. (1999) (Zhao and Liu 2012)

soft part, as expected, is subject to relatively larger deformation at low stress. Thirdly, the estimated γ_t values for the 11 clay rock samples under consideration range from 0.11 to 0.69 %, smaller than the typical porosity of Opalinus Clay rock (12–21 %) (Corkum and Martin 2007a). This difference again suggests that the soft part is only a small percentage of pore volume. These observations are consistent with those discussed in Sect. 3.2.2.

To further verify the stress-strain relationship, Zhao and Liu (2012) compare the theoretical results with data from triaxial compression tests for shale rock (Xu et al. 2006) and conglomerate rock (Hu and Liu 2004). These triaxial tests involve a cylindrical rock sample subject to a constant confining pressure σ_c (corresponding

Table 3.4 Fitted parameter values from the experiment data for Opalinus Clay rock (Zhao and Liu 2012)

Specimen	E_e (MPa)	γ_e	E_t (MPa)	γ_t
bra2-2a	2537.5	0.99310	0.3	0.00690
bra1-7a	2643.5	0.99490	0.7	0.00510
bra1-3a	2438.8	0.99640	1.2	0.00360
bra2-2b	2414.4	0.99817	1.0	0.00183
bra1-7b	2682.9	0.99892	0.7	0.00108
9949a	2080.0	0.99520	0.3	0.00480
9963	2448.8	0.99640	0.2	0.00360
9949b	2326.9	0.99703	0.5	0.00297
9972	2164.5	0.99793	0.6	0.00207
9957	2804.4	0.99844	0.6	0.00156
9984b	3345.1	0.99847	0.6	0.00153

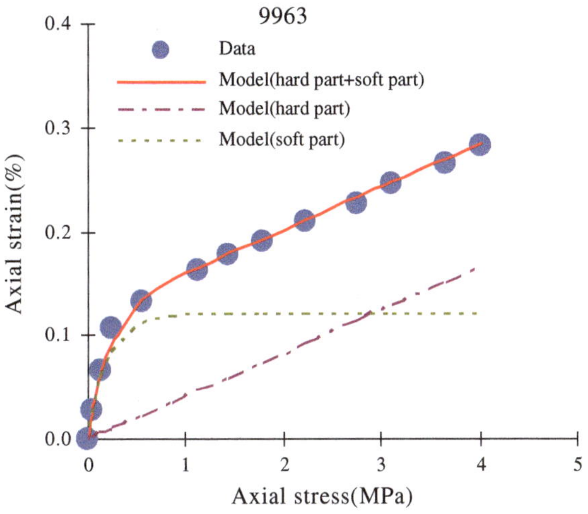

Fig. 3.15 Comparisons between the hard-part strain and the soft-part strain for specimen 9963 (Zhao and Liu 2012)

to axial strain ε_c) and then controlled increases in stress σ_1. In these tests, the measured relationship between deviatoric stress $(\sigma_1 - \sigma_c)$ and axial strain $(\varepsilon_1 - \varepsilon_c)$ are generally reported and used in the evaluation. When the confining pressure is constant, applying Eq. 3.49 yields:

$$\varepsilon_1 - \varepsilon_c = \frac{(3-\gamma_t)}{3E_e}(\sigma_1 - \sigma_c) + \frac{\gamma_t}{3}\exp\left(-\frac{\sigma_c}{E_t}\right)\left[1 - \exp\left(-\frac{\sigma_1 - \sigma_c}{E_t}\right)\right] \quad (3.53)$$

Figures 3.16 and 3.17 show the satisfactory matches of Eq. 3.53 with observed data from rock samples under triaxial compression conditions. The curve-fitted results indicate that the γ_t value ranges from 0.9 to 1.02 % for the conglomerate rock, and is 1.65 % for the shale rock. Because different rock samples are used for different stress conditions during the triaxial compression tests, some variations in

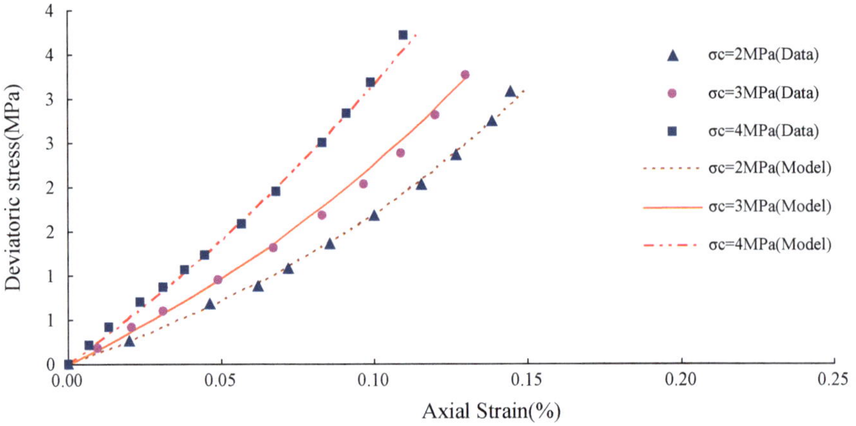

Fig. 3.16 Matches between results calculated from Eq. 3.53 and experimental data from triaxial compression tests on a conglomerate rock reported by Hu and Liu (2004) (Zhao and Liu 2012). The parameter σ_c is confining stress (pressure)

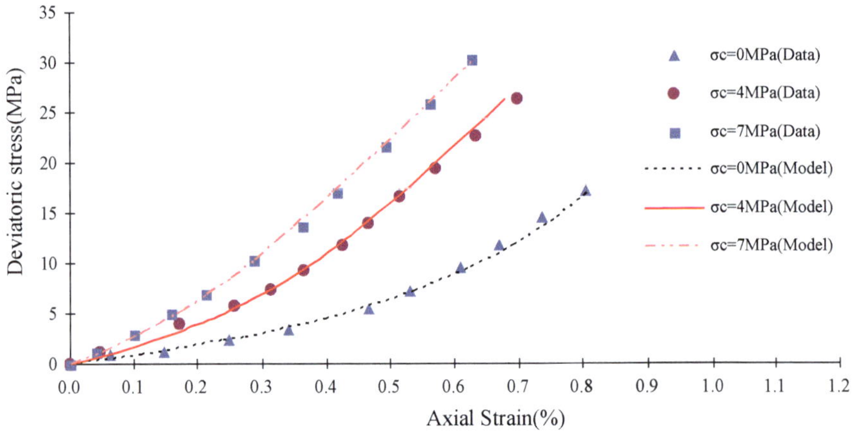

Fig. 3.17 Matches between results calculated from Eq. 3.53 and experimental data from triaxial compression tests on a shale rock reported in Xu et al. (2006) (Zhao and Liu 2012). The parameter σ_c is confining stress (pressure)

the fitted values for rock parameters are observed for a given rock type. The fitted parameter values are listed in Table 3.5.

To demonstrate the relative importance of the soft part under a triaxially stressed state, Fig. 3.18 shows the results of both the soft-part strain and the ratio of soft-part strain to the hard-part strain, *R*, as a function of axial stress for a shale sample. The curve describes the overall deformation behavior of the soft part, showing a significant initial increase in strain with stress and then a slower change later. The *R* decreases with increased confining pressure at a given deviatoric stress, and also

Table 3.5 Fitted parameter values from the experimental data (Zhao and Liu 2012)

Sample	Confining pressure (MPa)	E_e (MPa)	γ_e	E_t (MPa)	γ_t
Conglomerate rock	2	4670.0	0.9910	1.6	0.0090
	3	4670.0	0.9898	2.0	0.0102
	4	4670.0	0.9898	2.0	0.0102
Shale rock	0	6100.0	0.9835	5.0	0.0165
	4	6100.0	0.9835	5.0	0.0165
	7	6100.0	0.9835	5.0	0.0165

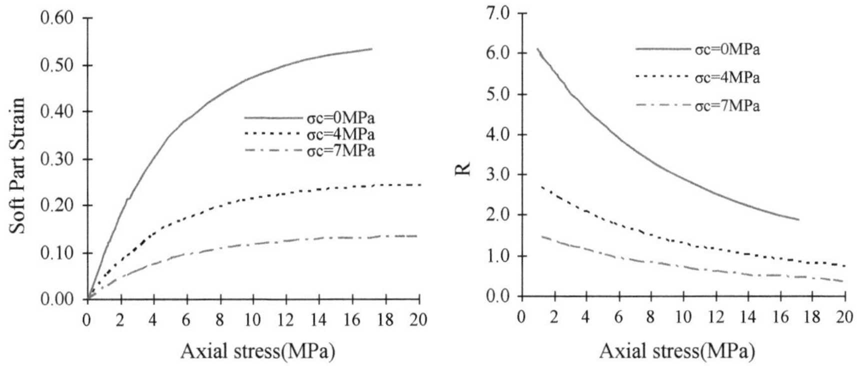

Fig. 3.18 Soft-part strain and R (the ratio of soft-part strain to the hard-part strain) as a function of axial stress at different confining pressures for a shale rock (Zhao and Liu 2012). The parameter σ_c is confining stress (pressure)

with increased axial stress at a given deviatoric stress. The conglomerate rock samples show the similar behavior.

In summary, the laboratory data of stress-strain relations, obtained under the anisotropic stress condition, are consistent with the theoretical development presented in this subsection (Zhao and Liu 2012).

3.2.4 TPHM-Based Constitutive Relationships for Anisotropic Stress Condition

The stress-strain relationships presented in Sect. 3.2.3 allow derivation of a variety of additional constitutive relationships between stress and mechanical/hydraulic properties for the anisotropic stress condition. This subsection presents the stress dependence of porosity, compressibility and shear modulus, as illustrative examples. The content of this subsection is based on the materials from Zhao and Liu (2012).

3.2.4.1 Rock Porosity

Following the procedure to derive Eq. 3.26, we assume that the soft part is a fraction of pore space. In this case, we rewrite Eq. 3.23 as

$$d\phi = \frac{dV^p}{V} = \frac{dV_e^p + dV_t}{V} \approx \frac{dV_e^p + dV_t}{V_0} \tag{3.54}$$

where again V is the bulk volume of rock and subscript p refers to pore space. Liu et al. (2009) indicated that for the purpose of calculating porosity, the total rock volume V could be approximated with the unstressed volume V_0, because their differences are small in practical applications.

For the hard part of the pore space, we have

$$\frac{dV_e^p}{V_0} = \frac{V_{0,e}^p}{V_0}\frac{dV_e^p}{V_{0,e}^p} = -(\phi_0 - \gamma_t)C_e(d\sigma_1 + d\sigma_2 + d\sigma_3) \tag{3.55}$$

To derive the above equation, we use the following relations:

$$\frac{V_{0,e}^p}{V_0} = \phi_0 - \gamma_t \tag{3.56}$$

and

$$C_e = -\frac{1}{3V_{0,e}^p}\frac{\partial V_e^p}{\partial \sigma_i} \quad (\mathrm{i} = 1, 2, 3) \tag{3.57}$$

where $V_{0,e}^p$ is hard part of the pore volume under unstressed conditions, ϕ_0 again is porosity under unstressed conditions, and C_e is the pore compressibility (and constant).

From Eq. 3.45 and its derivation procedure, it can be mathematically shown that the porosity change owing to the soft part, $\frac{dV_t}{V_0}$, is the same as the last three terms on the right-hand side of Eq 3.45. Thus, based on Eqs. 3.54 and 3.55, we have

$$\begin{aligned} d\phi &= -(\phi_0 - \gamma_t)C_e(d\sigma_1 + d\sigma_2 + d\sigma_3) \\ &- \frac{\gamma_t}{3E_t} \times \left[\exp\left(-\frac{\sigma_1}{E_t}\right)d\sigma_1 + \exp\left(-\frac{\sigma_2}{E_t}\right)d\sigma_2 + \exp\left(-\frac{\sigma_3}{E_t}\right)d\sigma_3\right] \end{aligned} \tag{3.58}$$

Using the condition that unstressed porosity is ϕ_0, we obtain the following (Zhao and Liu 2012)

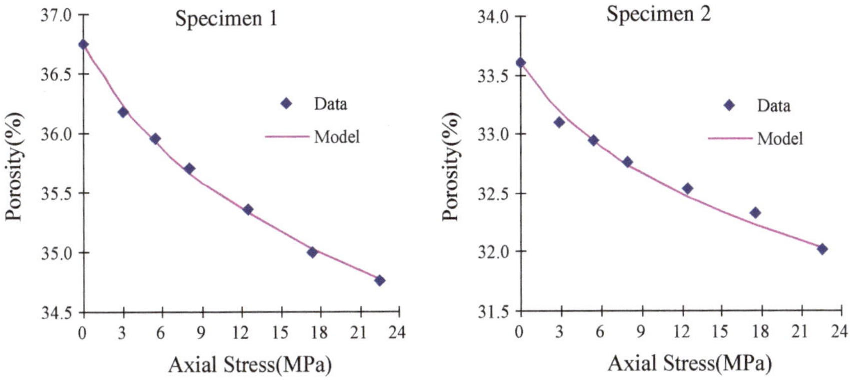

Fig. 3.19 Matches between results calculated from Eq. 3.59 and experimental data from uniaxial strain tests for two sand specimens reported in Peng and Zhang (2007) (Zhao and Liu 2012)

Table 3.6 Fitted parameter values from the experiment data of the sandstone samples (Zhao and Liu 2012)

Specimens	ϕ_0 (%)	C_e (10^{-4} MPa^{-1})	E_t (MPa)	γ_t
1	36.75	5.11	5.0	0.0180
2	33.60	4.44	5.0	0.0144

$$\begin{aligned}\phi &= \phi_0 - \gamma_t - (\phi_0 - \gamma_t)C_e(\sigma_1 + \sigma_2 + \sigma_3) \\ &\quad + \frac{\gamma_t}{3}\left[\exp\left(-\frac{\sigma_1}{E_t}\right) + \exp\left(-\frac{\sigma_2}{E_t}\right) + \exp\left(-\frac{\sigma_3}{E_t}\right)\right]\end{aligned} \tag{3.59}$$

Experimental results from uniaxial strain tests are used to verify the porosity-stress relation, or Eq. 3.59; relevant data are very limited for more complex stress conditions. Peng and Zhang (2007) reported a data set of porosity (as a function of axial stress) under uniaxial strain conditions for two sandstone specimens from 1000 m below the sea floor. Satisfactory matches between results calculated from Eq. 3.59 and the porosity data are shown in Fig. 3.19. The fitted parameter values are given in Table 3.6.

3.2.4.2 Bulk Compressibility

The relationship between bulk compressibility and stress is given in Sect. 3.2.2.1 for the isotropic-stress condition. A more general relationship is given in this subsection. Here, we define the bulk compressibility (associated with a principal stress σ_i) by

$$C_i = \frac{\partial \varepsilon_v}{\partial \sigma_i} \quad (\mathrm{i} = 1, 2, 3) \tag{3.60}$$

Based on the above equation and Eq. 3.45, the compressibility can be readily determined as (Zhao and Liu 2012)

$$C_i = \frac{(3-\gamma_t)(1-2\gamma)}{3E_e} + \frac{\gamma_t}{3E_t}\exp\left(-\frac{\sigma_i}{E_t}\right) \quad (\mathrm{i} = 1, 2, 3) \tag{3.61}$$

Morgenstern and Tamuly Phukan (1969) investigated the relationship between the compressibility and stress for Bunter sandstone. Results from unconfined compression tests of Morgenstern and Tamuly Phukan (1969) were employed to verify the compressibility-stress relation (Eq 3.61). For unconfined compression tests, we need only to consider C_1, because $\sigma_2 = \sigma_3 = 0$. As shown in Fig. 3.20, the relation can satisfactorily match the data, further supporting the overall TPHM-

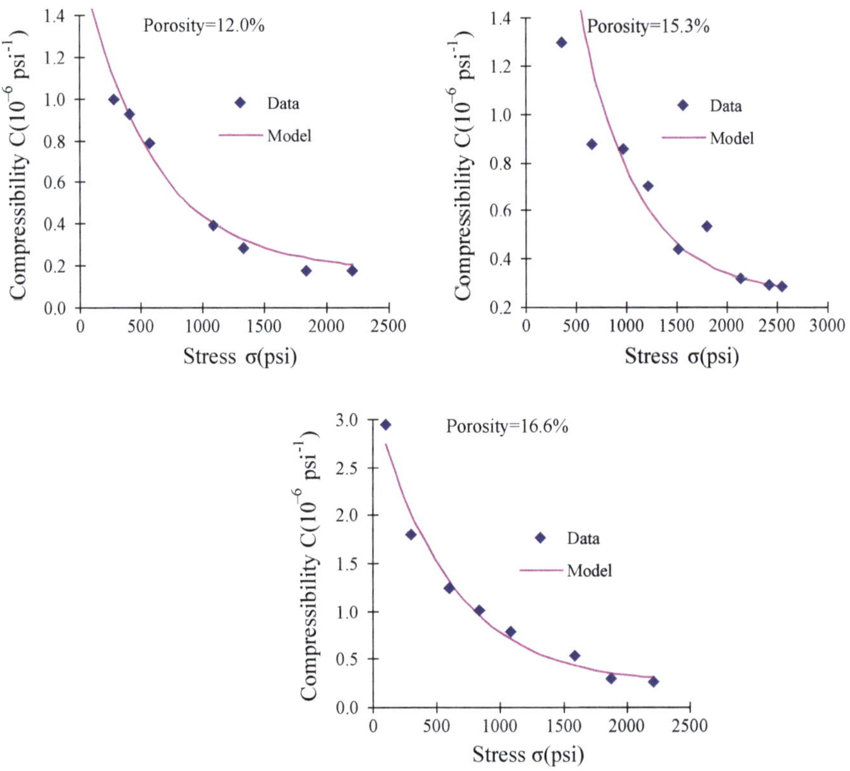

Fig. 3.20 Matches between the unconfined compression test data for Bunter sandstone from Morgenstern and Tamuly Phukan (1969) and the compressibility-stress relationship (Eq. 3.61) (Zhao and Liu 2012)

Table 3.7 Fitted parameter values from the experiment data for Bunter sandstone (Zhao and Liu 2012)

ϕ (%)	E_e (10^3 psi)	γ_e	E_t (10^3 psi)	γ_t
12.0	9914.000	0.99914	0.575	0.00086
15.3	7187.544	0.99827	0.575	0.00173
16.8	7187.544	0.99827	0.575	0.00173

based theoretical results developed for the anisotropic condition. The estimated parameter values for sandstone are presented in Table 3.7. The compressibility C, defined by Morgenstern and Tamuly Phukan (1969), is three times the compressibility given in Eq. 3.61. In Fig. 3.20, the former is used. Note that the value for E_t is generally consistent with those reported in Table 3.6 for the sandstone. However, the estimated γ_t values are much lower than those given in Table 3.6, which may be a result of the fact that rock porosity values in this case are much lower as well.

3.2.4.3 Shear Modulus

Shear modulus is an important parameter for various engineering projects and can be described as:

$$K_{xy} = \frac{d\tau_{xy}}{d\gamma_{xy}}, \; K_{xz} = \frac{d\tau_{xz}}{d\gamma_{xz}}, \; K_{yz} = \frac{d\tau_{yz}}{d\gamma_{yz}} \tag{3.62}$$

where τ and γ are shear stress and strain, respectively, and again x, y, z are spatial coordinates. Based on the above definitions and Eqs. 3.51 and 3.52, shear modulus K_{xy} can be easily determined as

$$K_{xy} = \frac{1}{\frac{2(3-\gamma_t)(1+v)}{3E_e} + \frac{2\gamma_t}{3E_t}\left[\exp\left(-\frac{\sigma_1}{E_t}\right)\frac{d\sigma_1}{d\tau_{xy}}l_1m_1 + \exp\left(-\frac{\sigma_2}{E_t}\right)\frac{d\sigma_2}{d\tau_{xy}}l_2m_2 + \exp\left(-\frac{\sigma_3}{E_t}\right)\frac{d\sigma_3}{d\tau_{xy}}l_3m_3\right]} \tag{3.63}$$

Under the stress condition of $\sigma_2 = \sigma_3$, one can get

$$K_{xy} = \frac{d\sigma_1 - d\sigma_2}{\frac{2(3-\gamma_t)(1+v)(d\sigma_1 - d\sigma_2)}{3E_e} + \frac{2\gamma_t}{3E_t}\left[\exp\left(-\frac{\sigma_1}{E_t}\right)d\sigma_1 - \exp\left(-\frac{\sigma_2}{E_t}\right)d\sigma_2\right]} \tag{3.64}$$

$$K_{xy} = K_{yz} = K_{xz} \tag{3.65}$$

Stress-dependent data for shear modulus are relatively limited in the literature. Thus, we use results calculated from Eq. 3.64 (with estimated parameters from Table 3.4) to demonstrate the stress-dependent behavior of shear modulus in Fig. 3.21. No comparison is made with experimental observations. However, note that Eq. 3.64 is the direct result of a mathematical transformation of Eq. 3.36. The

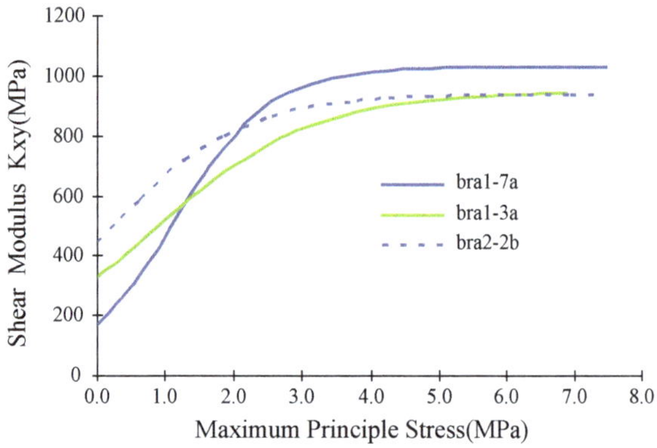

Fig. 3.21 Shear modulus as a function of σ_1 for the three rock samples with parameter values given in Table 3.4 (Zhao and Liu 2012)

validation of Eq. 3.36, discussed above, is equivalent to that of Eq. 3.65. For simplicity, we set $\sigma_2 = \sigma_3 = 0$ in Fig. 3.21, which shows the strong stress dependence of shear modulus at low stress values.

3.2.5 Implementation of the TPHM in a Geomechanical Simulator

The TPHM and the related constitutive relations can be implemented into a geomechanical simulator for practical applications. Incorporation of them into an existing geomechanical simulator may be a convenient choice in many cases. As an example, this subsection briefly discusses how the related constitutive relations are incorporated into FLAC3D (Itasca Consulting Group 2005), a geomechanical simulator that has been widely used in the fields of geotechnical, geomechanical, civil, and mining engineering, while a more detailed presentation of the implementation procedure can be found in Li et al. (2014).

FLAC3D uses dynamic equations of motion in its explicit, time-marching scheme. The solution of solid body problems in FLAC3D invokes the equations of motion (Newton's law of motion), constitutive relationships, and boundary conditions. When running the FLAC3D code in its mechanical or thermomechanical configuration mode, the code solves the following equation of motion in an iterative manner with the stress–strain relationship:

$$\nabla \cdot \boldsymbol{\sigma} + \rho_m \mathbf{g} = \rho_m \frac{d\mathbf{v}}{dt} \tag{3.66}$$

where ρ_m is average density of the rock mass and $\mathbf{v}$ is solid velocity with respect to a fixed system. The incremental stress and strain during a time step is governed by various elastic or elasto-plastic constitutive relationships that can be written in a general form as:

$$\Delta\sigma' = \mathbf{H}(\sigma', \dot{\varepsilon}\Delta t) \tag{3.67}$$

in which $\mathbf{H}$ contains given material functions, $\dot{\varepsilon}$ is the infinitesimal strain-rate tensor, and Δt is a time increment. The constitutive relationships in the above equation are applied to the effective stress calculated as:

$$\boldsymbol{\sigma} = \boldsymbol{\sigma}_{\mathrm{T}} + \boldsymbol{I}\alpha_{\mathrm{B}}P \tag{3.68}$$

where α_B is Biot's effective stress parameter (Biot 1941), P is pore pressure, $\boldsymbol{I}$ is unit tensor and $\boldsymbol{\sigma}$ and $\boldsymbol{\sigma_T}$ are effective stress and total stress tensors, respectively. One useful feature of FLAC3D is that users can modify the existing constitutive models in the simulator or create their own constitutive models as dynamic-linked libraries (DLLs).

All constitutive models in FLAC3D share the same incremental numerical algorithm. Given the stress state at time t and the strain increment for a time step Δt, the purpose of the constitutive models is to determine the corresponding stress increment and the new stress state at time $t + \Delta t$. When plastic deformations are involved, only the elastic part of the strain increment will contribute to the stress increment. In this case, a correction must be made to the elastic stress increment (as computed from the total strain increment) in order to obtain the actual stress state for the new time step.

As a stress-strain relationship, the TPHM is implemented into FLAC3D by modifying the Mohr–Coulomb model; the linear elastic portion of the Mohr–Coulomb model is replaced by the TPHM. To do so, an elastic guess (σ_{ij}^{I}) at time $t + \Delta t$ is first computed by application of the TPHM to the given strain increments $\Delta\varepsilon_{ij}$, for i, j = 1, 2, 3. New stress values at time $t + \Delta t$, elastic guesses $\left(\sigma_{ij}^{I}\right)$, are then obtained from the relationship

$$\sigma_{ij}^{I} = \sigma_{ij}^{0} + \Delta\sigma_{ij}^{TPHM} \tag{3.69}$$

where σ_{ij}^{0} are the initial stresses at time t. If the new stresses violate the yield criterions, plastic deformation takes place and needs to be considered. Then the new stress components are corrected by using the plastic flow rule to ensure that they lie on the yield surface in stress space. If the point representing new stresses in the stress space is located below failure envelope, no plastic flow takes place for this step and stresses obtained from Eq. 3.69 are acceptable. In Eq. 3.69, the $\Delta\sigma_{ij}^{TPHM}$ can be directly obtained from the following equation:

$$\left.\begin{aligned}
d\varepsilon_x &= \frac{(3-\gamma_t)}{3E_e}\left[d\sigma_x - v\left(d\sigma_y + d\sigma_z\right)\right] + d\varepsilon_{xt} \\
d\varepsilon_y &= \frac{(3-\gamma_t)}{3E_e}\left[d\sigma_y - v\left(d\sigma_x + d\sigma_z\right)\right] + d\varepsilon_{yt} \\
d\varepsilon_z &= \frac{(3-\gamma_t)}{3E_e}\left[d\sigma_z - v\left(d\sigma_x + d\sigma_y\right)\right] + d\varepsilon_{zt} \\
d\gamma_{xy} &= \frac{2(3-\gamma_t)(1+v)}{3E_e} d\tau_{xy} + d\gamma_{xyt} \\
d\gamma_{xz} &= \frac{2(3-\gamma_t)(1+v)}{3E_e} d\tau_{xz} + d\gamma_{xzt} \\
d\gamma_{yz} &= \frac{2(3-\gamma_t)(1+v)}{3E_e} d\tau_{yz} + d\gamma_{yzt}
\end{aligned}\right\} \tag{3.70}$$

Equation 3.70 is an incremental formulation of Eqs. 3.50 and 3.51. In Eq. 3.70, the strain increments $d\varepsilon_x$, $d\varepsilon_y$, $d\varepsilon_z$, $d\gamma_{xy}$, $d\gamma_{xz}$, $d\gamma_{yz}$ are known values, i.e., $\Delta\varepsilon_{ij}$, while the right-hand-side items $d\varepsilon_{xt}$, $d\varepsilon_{yt}$, $d\varepsilon_{zt}$, $d\gamma_{xyt}$, $d\gamma_{xzt}$, $d\gamma_{yzt}$ are the strain increments for the soft part and can be expressed as

$$\left.\begin{aligned}
d\varepsilon_{xt} &= \frac{\gamma_t}{3E_t}\left[\exp\left(-\frac{\sigma_1}{E_t}\right)l_1^2 d\sigma_1 + \exp\left(-\frac{\sigma_2}{E_t}\right)l_2^2 d\sigma_2 + \exp\left(-\frac{\sigma_3}{E_t}\right)l_3^2 d\sigma_3\right] \\
d\varepsilon_{yt} &= \frac{\gamma_t}{3E_t}\left[\exp\left(-\frac{\sigma_1}{E_t}\right)m_1^2 d\sigma_1 + \exp\left(-\frac{\sigma_2}{E_t}\right)m_2^2 d\sigma_2 + \exp\left(-\frac{\sigma_3}{E_t}\right)m_3^2 d\sigma_3\right] \\
d\varepsilon_{zt} &= \frac{\gamma_t}{3}\left[\exp\left(-\frac{\sigma_1}{E_t}\right)n_1^2 d\sigma_1 + \exp\left(-\frac{\sigma_2}{E_t}\right)n_2^2 d\sigma_2 + \exp\left(-\frac{\sigma_3}{E_t}\right)n_3^2 d\sigma_3\right] \\
d\gamma_{xyt} &= \frac{2\gamma_t}{3E_t}\left[\exp\left(-\frac{\sigma_2}{E_t}\right)d\sigma_2 - \exp\left(-\frac{\sigma_1}{E_t}\right)d\sigma_1\right]l_2 m_2 + \frac{2\gamma_t}{3}\left[\exp\left(-\frac{\sigma_3}{E_t}\right)d\sigma_3 - \exp\left(-\frac{\sigma_1}{E_t}\right)d\sigma_1\right]l_3 m_3 \\
d\gamma_{xzt} &= \frac{2\gamma_t}{3E_t}\left[\exp\left(-\frac{\sigma_2}{E_t}\right)d\sigma_2 - \exp\left(-\frac{\sigma_1}{E_t}\right)d\sigma_1\right]l_2 n_2 + \frac{2\gamma_t}{3}\left[\exp\left(-\frac{\sigma_3}{E_t}\right)d\sigma_3 - \exp\left(-\frac{\sigma_1}{E_t}\right)d\sigma_1\right]l_3 n_3 \\
d\gamma_{yzt} &= \frac{2\gamma_t}{3E_t}\left[\exp\left(-\frac{\sigma_2}{E_t}\right)d\sigma_2 - \exp\left(-\frac{\sigma_1}{E_t}\right)d\sigma_1\right]m_2 n_2 + \frac{2\gamma_t}{3}\left[\exp\left(-\frac{\sigma_3}{E_t}\right)d\sigma_3 - \exp\left(-\frac{\sigma_1}{E_t}\right)d\sigma_1\right]m_3 n_3
\end{aligned}\right\} \tag{3.71}$$

In Eq. 3.71, the principal stresses σ_1, σ_2 and σ_3 (not their increments) are approximated with the stresses σ_{ij}^0 at time t based on the stress transformations in three dimensions (Poulos and Davis 1974; Jaeger et al. 2007). An algorithm function for calculating σ_1, σ_2 and σ_3 has been provided by the module of the C++ source code for the Mohr–Coulomb model in FLAC3D. The principal stress increments $d\sigma_1$, $d\sigma_2$, and $d\sigma_3$ in Eq. 3.71 can be expressed as

$$\left.\begin{aligned}
d\sigma_1 &= d\sigma_x l_1^2 + d\sigma_y m_1^2 + d\sigma_z n_1^2 + 2d\tau_{xy}l_1 m_1 + 2d\tau_{xz}l_1 n_1 + 2d\tau_{yz}m_1 n_1 \\
d\sigma_2 &= d\sigma_x l_2^2 + d\sigma_y m_2^2 + d\sigma_z n_2^2 + 2d\tau_{xy}l_2 m_2 + 2d\tau_{xz}l_2 n_2 + 2d\tau_{yz}m_2 n_2 \\
d\sigma_3 &= d\sigma_x l_3^2 + d\sigma_y m_3^2 + d\sigma_z n_3^2 + 2d\tau_{xy}l_3 m_3 + 2d\tau_{xz}l_3 n_3 + 2d\tau_{yz}m_3 n_3
\end{aligned}\right\} \tag{3.72}$$

where l_i, m_i, and n_i ($i = 1, 2, 3$) can be calculated by Eq. 3.52 using stresses at time t.

After inserting Eqs. 3.71 and 3.72 into Eq. 3.70, Eq. 3.70 becomes a system of linear equations with unknowns $d\sigma_x$, $d\sigma_y$, $d\sigma_z$, $d\tau_{xy}$, $d\tau_{xz}$ and $d\tau_{yz}$ that correspond to $\Delta\sigma_{ij}^{TPHM}$ in Eq. 3.69. Thus, one can obtain $\Delta\sigma_{ij}^{TPHM}$ by solving the system of linear equations.

Note that the TPHM is only applicable to compression conditions. The soft part, just like a pore or fracture, cannot sustain tensions, and thus the conventional Hook's law is used in this case. In other words, if one of the principal stresses is in tension, the strain for the soft part in the same direction as the principal stress will be omitted. The computational scheme implemented into FLAC3D is shown in Fig. 3.22.

3.3 Fracture Deformation and Properties

Fractures, existing at different spatial scales, are the common geological structures in the earth's upper crust. Fracture deformation has an important effect on fluid flow in a discontinuous rock mass under changing stress conditions (Bandis et al. 1983). Specifically, fluid flow takes place predominantly through fractures in a fractured rock rather than the rock matrix, and the permeability of a rock depends strongly on fracture apertures. An improved understanding of fracture deformation is needed for a number of practical applications, such as hazardous waste disposal, coal mining, oil recovery from fractured reservoirs, natural gas recovery from shale formations through hydraulic fracturing, CO_2 geological sequestration, and geothermal energy extraction. Many experimental and theoretical studies have thus been conducted focusing on the mechanical properties of fractures and their relationship with hydraulic properties (Brown and Scholz 1986).

In Sect. 3.2, we present the basic concepts of the TPHM and its applications to rock matrix (or porous media). In this section, we will further demonstrate that the TPHM is also applicable to fractures. Fracture deformation can occur under compression and/or shearing. The focus of this section is on fracture deformation driven by the normal stress because it is directly relevant to the TPHM. The readers are referred to Wei et al. (2013) for a recent review of the progress in determining fracture deformation resulting from shearing. The content of this section is based on the materials from Liu et al. (2011, 2013) and Liu and Rutqvist (2010).

3.3.1 Normal-Stress Dependence of Fracture Hydraulic Properties

This subsection presents the TPHM-based relationships between normal stress and fracture hydraulic properties; the latter are also strong functions of shearing (e.g., Bandis et al. 1983). The content of this subsection is based on the study of Liu et al. (2013).

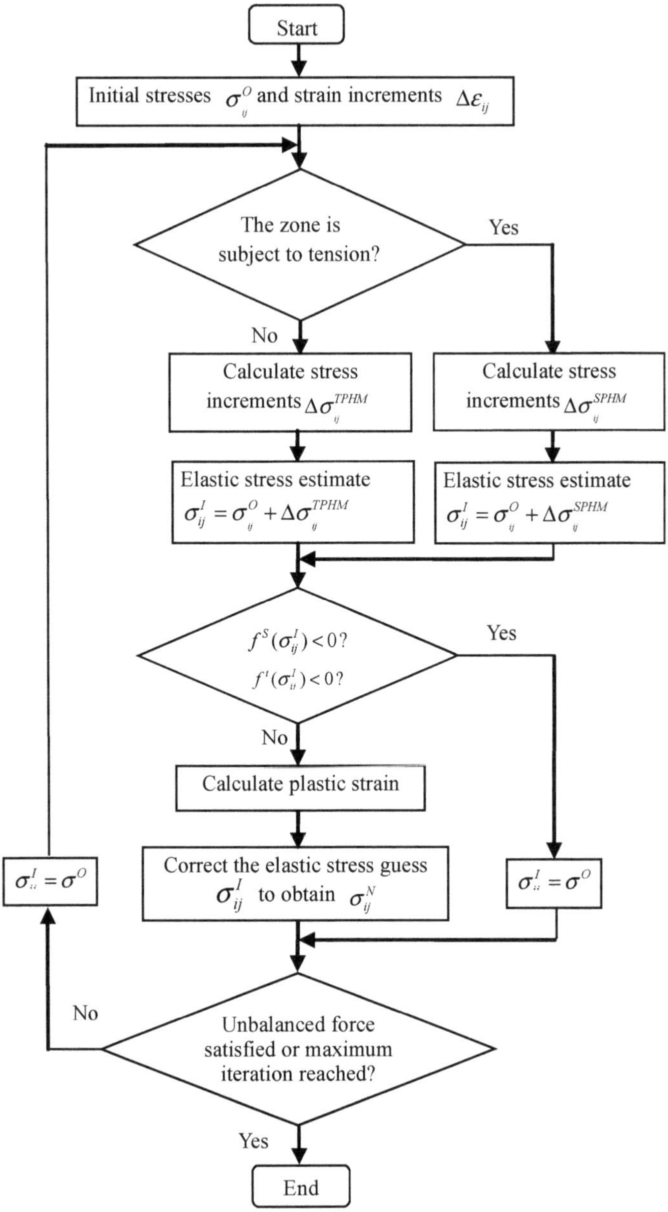

Fig. 3.22 Computational scheme for the numerical implementation of the TPHM into FLAC3D (Li et al. 2014) (reproduced by permission of Elsevier)

The stress dependence of fracture hydraulic properties (e.g., permeability, aperture or closure) has been investigated by a number of researchers. Note that fracture permeability and fracture aperture (or closure) are closely related through the cubic law and its variations (e.g., Zimmermann and Bodvarsson 1996). Goodman (1974) proposed an empirical hyperbolic relationship between normal stress and fracture closure that was later modified by Bandis et al. (1983) to better fit observations. Other models, based on Hertzian theory, have also been developed to describe the nonlinear stress-deformation behavior (McDermott and Kolditz 2006). These models suggested that the observed nonlinear behavior could be attributed to the increasing contact areas as the normal stress increases. For example, Brown and Scholz (1985, 1986) used Hertzian theory to study fracture closure as a function of normal stress. Based on Hertzian theory of deformation of spheres, Gangi (1978) developed a relationship for stress-dependent fracture permeability. Most recently, Liu et al. (2009) proposed the TPHM for describing the relationship between stress and elastic strain, which has been discussed in previous sections of this chapter. The main idea of the TPHM is to capture heterogeneous deformation processes at a macroscopic scale, resulting from the existence of heterogeneity of rock mass, by conceptualizing the rock mass (or a fracture) into two parts with different mechanical properties. While the TPHM has been validated for different rock types, evaluation of its validity for fractures is limited (Liu et al. 2009, 2011; Zhao and Liu 2012). This subsection gives a more comprehensive evaluation of the TPHM using data on fracture deformation gathered from the literature. The normal stress dependence of fracture two-phase flow properties will also be discussed.

3.3.1.1 Fracture Aperture or Closure Under Different Normal Stresses

Fracture aperture is an important parameter for both mechanical and hydraulic processes within a fractured rock. In this subsection, a relationship, based on the TPHM, is developed for the dependence of fracture aperture (or closure) on the normal stress (Liu et al. 2013). Also note that the term stress here refers to effective stress to take account of the effect of pore-liquid pressure.

Following Sect. 3.2, the subscripts 0, e, and t are used again to denote the unstressed state, the hard part and the soft part, respectively. Consider a fracture to be subject to a normal stress σ_n and divide fracture space into hard and soft parts along the direction normal to the fracture plane. Then, the volumetrically-averaged fracture aperture (b) is given by:

$$b_0 = b_{0,e} + b_{0,t} \tag{3.73}$$

under the unstressed condition, and

$$b = b_e + b_t \tag{3.74}$$

under the stressed condition. Hooke's Law for the two parts can be expressed by

$$d\sigma_n = -K_{e,f}\frac{db_e}{b_{0,e}} \tag{3.75}$$

$$d\sigma_n = -K_{t,f}\frac{db_t}{b_t} \tag{3.76}$$

where K is the fracture Young's modulus and subscript f refers to fracture. The stress in the two equations above refers to far-field normal stress, rather than local stress.

Combining Eqs. 3.73–3.76 gives

$$db = db_e + db_t = -b_{0,e}\frac{d\sigma_n}{K_{e,f}} - b_t\frac{d\sigma_n}{K_{t,f}} \tag{3.77}$$

Integrating the above equation and using the following relationship obtained from Eq. 3.75:

$$b_t = b_{0,t}\exp\left(-\frac{\sigma_n}{K_{t,f}}\right) \tag{3.78}$$

Liu et al. (2013) obtained

$$b = b_{0,e}\left(1 - \frac{\sigma_n}{K_{e,f}}\right) + b_{0,t}\exp\left(-\frac{\sigma_n}{K_{t,f}}\right) \tag{3.79}$$

Note that the derivation of Eq. 3.79 uses the condition that for $\sigma_n = 0$, $b_e = b_{0,e}$ and $b_t = b_{0,t}$.

In Eq. 3.79, the stress-dependent behavior of fracture aperture is controlled by the second term at a low stress and the first term at a high stress. However, in many laboratory tests, fracture closure, rather than aperture, is measured. The fracture closure (χ) as a function of normal stress can be derived from Eq. 3.79 as follows (Liu et al. 2013):

$$\chi = b_0 - b = b_{0,e}\left(\frac{\sigma_n}{K_{e,f}}\right) + b_{0,t}\left[1 - \exp\left(-\frac{\sigma_n}{K_{t,f}}\right)\right] \tag{3.80}$$

Equations 3.79 and 3.80 give relationships between fracture aperture, closure and normal stress.

In the above equations, fracture aperture b is a mechanical aperture, rather than the hydraulic fracture aperture that determines fracture permeability through the cubic law (Witherspoon et al. 1980; Olsson and Barton 2001). These two apertures are generally different, except for the case of smooth fractures. Barton et al. (1985) and Zimmerman and Bodvarsson (1996) proposed mathematical expressions to relate these two apertures that are functions of fracture roughness and contact area. When information on fracture roughness and contact area is available or can be estimated, the stress-dependence of fracture permeability can be estimated through the cubic law and one of these relations between mechanical and hydraulic apertures. A simpler approach is to use the ratio of mechanical apertures at different stresses to approximate the corresponding ratio of hydraulic apertures for a given fracture. In this case, the cubic law leads to (Liu et al. 2013):

$$\frac{k}{k_0} = \left(\frac{b}{b_0}\right)^3 \tag{3.81}$$

where k is the fracture permeability corresponding to fracture aperture b, and again subscript 0 refers to the zero-stress condition. This treatment seems to be supported by several studies. For example, Elliott and Brown (1988) experimentally show that flow rate through a fracture is approximately proportional to the cube of mechanical aperture for a range of aperture values (15–21 μm). It is obvious from the cubic law that their results essentially indicate the proportionality of mechanical aperture to hydraulic aperture under the corresponding test conditions. Liu et al. (2009, 2011) successfully match two data sets of fracture permeability as a function of stress using Eq. 3.81. Therefore, the above equation can be used as a first-order approximation for practical applications.

It is useful to further emphasize that the TPHM has two key elements. Firstly, it hypothesizes that in Hooke's Law, true strain, rather than engineering strain, should be used. However, because the hard part is subject to a small deformation, engineering and true strains are practically identical for the hard part. In this case, the engineering strain, for the mathematical simplicity, is used to describe the mechanical deformation of the hard part. Secondly, the TPHM conceptualizes the rock mass (or a fracture) to consist of two parts, hard and soft. This is largely motivated by the well-known heterogeneity of subsurface material. The existence of the soft part in porous media, in terms of micro-cracks and pore space near the grain contacts, has been discussed in previous investigations (e.g., Mavko and Jizba 1991) and in Sect. 3.2. The soft and hard parts for fracture apertures are those corresponding to relatively small and large contact areas within fractures, respectively. Nevertheless, the TPHM, as a macroscopic model that is not derived from micro-mechanics, should be ultimately evaluated by how consistent it is with a large range of observations.

To evaluate the usefulness of Eqs. 3.79 and 3.80, we examine whether these equations can satisfactorily match experimental observations from different sources (Liu et al. 2013). Since most laboratory data are for fracture closures, the focus of the evaluation is on Eq. 3.80. A relatively comprehensive literature survey of the closure-stress data was performed. The selection of the data sets to be used for the comparisons is based on several considerations. Firstly, the data set and its measurement procedure must be well documented, such that there is a good understanding of the data set and the possible errors involved in the measurements. Secondly, as suggested by Barton et al. (1985), the closure-stress measurements may be representative of in situ fracture behavior from the third or fourth loading cycles. Therefore, the analysis is focused on the data obtained from these cycles, or data corresponding to the largest cycle numbers when measurements from the third or fourth loading cycles are not available. Table 3.8 presents information on the data sets used in this study. Readers are referred to the original data sources listed in Table 3.8 for the details of data-collection procedures. The literature survey is by no means exhaustive, but Table 3.8 should include typical data sets for closure-stress measurements available in the public literature.

As examples, Fig. 3.23 shows comparisons between theoretical results from Eq. 3.80 and measured data from two experiments. The comparisons are typical for all the other experiments. To avoid non-uniqueness in the parameter estimation, values for parameters in Eq. 3.80, for a given data set, are determined in the manner similar to that used for determining parameters characterizing stress-dependent rock porosity. In a closure-stress plot (Fig. 3.23), data at high stress can be approximately represented by a linear relationship. As indicated by Eq. 3.80, the slope of the linear relationship is $b_{0,e}/K_{e,f}$ and its intersection with the closure axis is $b_{0,t}$. Then $K_{t,f}$ can be estimated from a data point at a low stress. In order to estimate $b_{0,e}$, additional information on either $K_{e,f}$ or total aperture b_0 under the unstressed condition is needed. In some cases, the information is not available (Table 3.8). This, however, does not impact on the evaluation of the TPHM, because Eq. 3.80 only requires the ratio $b_{0,e}/K_{e,f}$, rather than $b_{0,e}$, to calculate a fracture closure.

Figure 3.23 and Table 3.8 indicate that the TPHM adequately represents experimental observations. As expected, significant portions of the fracture aperture (or closure) are soft and characterized by high nonlinearity, as shown in Fig. 3.23, and the modulus for the soft part is significantly lower than that of the hard part. Note that for most data sets collected for different rock types and by different researchers (Table 3.8), the values for the correlation coefficient of curve fitting (R^2) are above 0.95.

It is important to mention that the TPHM, as previously discussed, is a macroscopic model that deals with mechanisms of micro-mechanics in a phenomenological manner. Its validity is demonstrated by its consistence with a number of data sets (Table 3.8). The fitted values for $K_{t,f}$ are relatively small, because they characterize fracture deformation at low normal stress corresponding to small contact areas within fractures. The values for $K_{e,f}$ are much larger, because $K_{e,f}$ characterizes fracture deformation corresponding to relatively stabilized contact

Table 3.8 Values for fitting parameters from experimental data (Liu et al. 2013)

Rock type	Initial aperture	Sample # (loading cycle)	Fitting parameters (as defined in the text)						Reference
			$b_{0,e}$ (mm)	$K_{e,f}$ (MPa)	$b_{0,e}/K_{e,f}$ (mm/MPa)	$b_{0,t}$ (mm)	$K_{t,f}$ (MPa)	R^2	
Granite	0.016 mm	Sample 1	0.00748	4179.3	0.000120	0.00852	3.81679	0.985	Indraratna et al. (1999)
	0.006 mm	Sample 2	0.00315	15181	3.29359E−05	0.00285	3.81679	0.985	
	0.0045 mm	Sample 3	0.00033	16160	3.09406E−05	0.00417	1.79856	0.947	
Diorite	0.38 mm	Dr-1	0.128	39.635	0.003229	0.25200	2.31922	0.985	Malama and Kulatilake (2003)
Granodiorite rock	0.505 mm	Grd-1	0.046	36.508	0.001260	0.45913	2.07031	0.975	
	0.481 mm	Grd-2	0.042	39.628	0.001060	0.43949	2.38408	0.998	
	0.683 mm	Grd-3	0.064	25.106	0.002549	0.61854	2.00781	0.990	
Slate	≤0.1 mm	Fresh-slate cleavage (3)	0.0750	2309.814	3.24749E−05	0.02499	7.35294	0.991	Bandis et al. (1983)
Dolerite	0.15 mm	Fresh-dolerite joint (3)	0.1179	4678.191	2.52003E−05	0.03211	5.84795	0.987	
Limestone	0.2 mm	Fresh-limestone bedding (3)	0.1875	8386.160	2.23624E−05	0.01247	3.96825	0.999	
Slate	0.5 mm	Weathered-slate cleavage (3)	0.4303	1506.220	0.000286	0.06965	5.52486	0.996	
Limestone	0.5 mm	Weathered-limestone joint (3)	0.4336	2004.390	0.000216	0.06644	9.70874	0.977	
Siltstone	0.6 mm	Weathered-siltstone bedding (3)	0.5286	2374.240	0.000223	0.07137	6.62252	0.996	
Fused-silica glass	Max = 113 μm	CGL965 (cycle 4, 5)	0.1024	657.89	0.000156	0.01060	0.69686	0.928	Brown and Scholz (1985)
Fused-silica glass	Max = 37.8 μm	CGL0394	0.0304	377.39	8.05533E−05	0.00735	0.87184	0.969	
Fused-silica glass	Max = 28 μm	CGL0234	0.0204	90.054	0.000227	0.00757	0.72727	0.988	

(continued)

Table 3.8 (continued)

Cheshire quartzite	Max = 35.9 μm	CCQ0073	0.028	557.34	5.02386E−05	0.00793	6.36943	0.960	Brown and Scholz (1986)
Cheshire quartzite	Max = 74 μm	CCQ0102 (cycle 3)	0.0657	2204.2	2.98067E−05	0.00828	6.32911	0.927	
Marble	Max = 37.9 μm	CSM0042 (cycle 3)	0.0299	156.04	0.000192	0.00798	0.94518	0.951	
Carnmenellis granite	–	Sample 1	–	–	0.001453	0.06019	2.18341	0.955	Elliott and Brown (1988)
–	–	Mated	–	–	0.000591	0.10327	1.5674	0.997	Goodman (1976)
		Unmated			0.001979	0.34471	3.63636	0.992	
Granite	–	–	–	–	0.000569	0.10939	0.54975	0.988	Iwai (1976)
Kikuma granodiorite	262 μm	NKGD (cycle 2)	0.12	148.16	0.000810	0.14202	0.90416	0.903	Iwano (1995)
Kikuma granodiorite	627 μm	TKGD (cycle 2)	0.4511	272.84	0.001653	0.17588	3.11526	0.959	
Kikuma granodiorite	201 μm	SKGD (cycle 2)	0.0842	230.94	0.000365	0.11678	1.35685	0.924	
Inada granite	650 μm	TIGN (cycle 2)	0.4464	743.5	0.000600	0.20361	2.43309	0.905	
Chichibu schist	283 μm	TCSH (cycle 2)	0.1626	437.63	0.000372	0.11945	0.96993	0.984	
Kimachi sandstone	396 μm	TKSS (cycle 2)	0.0837	36.668	0.002283	0.31226	0.80645	0.882	
Carnmenellis granite	–	NJ1	–	–	0.002053	0.04239	0.38183	0.952	Zhao and Brown (1992)
		NJ2 (180 °C)			0.003940	0.04854	0.87336	0.967	
		NJ2 (200 °C)			0.003762	0.08679	0.48309	0.943	
		NJ6			0.007062	0.09304	0.77942	0.984	
Artificial tension fractures	–	EF1			0.001617	0.0536	1.14943	0.993	
		EF3			0.011696	0.01032	1.01937	0.981	
		EF7 (2nd loading-120 °C)			0.009166	0.0644	0.8244	0.997	

(continued)

Table 3.8 (continued)

Charcoal black granite	–	Sample 1 (cycle 3)	–	–	0.000246	0.02448	4.29185	0.992	Raven and Gale (1985)
		Sample 2 (cycle 3)			0.000623	0.06751	3.10559	0.923	
		Sample 3 (cycle 3)			0.001226	0.09352	1.14286	0.948	
		Sample 5 (cycle 3)			0.002049	0.07512	1.51976	0.990	
Inada granite	0.526 mm	IG10	0.438	219	0.002	0.088	0.83542	0.982	Matsuki et al. (2001)
	0.589 mm	IG15	0.5128	410.2	0.001250	0.07625	1.80505	0.965	
	0.39 mm	IG20	0.3386	176.07	0.001923	0.05141	1.83486	0.998	
	0.697 mm	IG25	0.61	586.54	0.001040	0.087	1.443	0.949	
Granite block	–	Sample 1	–	–	0.000599	0.085	1.28866	0.994	Sharifzadeh et al. (2008)
Quartz monzonite (Stripa granite)	–	E30 (cycle 2)	–	–	4.76281E−05	0.00493	12.1951	0.994	Pyrak-Nolte et al. (1987)
		E32			4.70057E−05	0.00151	3.77358	0.993	
		E35			0.000287	0.00341	2.89017	0.994	
Granite samples	–	GO3	–	–	0.000799	0.0157	10.6383	0.995	Chen et al. (2000)
Granodiorite	1001 μm	Load cycle 2	0.8711	403.07	0.002161	0.12993	2.88184	0.991	Schrauf and Evans (1986)
	600 μm	Uniform flow field	0.4968	170	0.002922	0.10319	0.5848	0.974	
	760 μm	Radial flow field	0.5961	168.55	0.003537	0.16386	1.54799	0.941	
Westerley (Rhode Island) granite	0.11–0.21 mm	Sample 1, mated	–	–	0.000125	0.1	17.5439	0.964	Durham and Bonner (1994)
	0.59 mm	Sample 1, offset			0.0005	0.27	16.9492	0.968	
Austin chalk	–	2nd Cycle	–	–	0.006351	0.04623	1.41243	0.996	Olsson and Brown (1993)
		Offset 1 mm			0.006774	0.12593	1.54321	0.998	
		Offset 3 mm			0.009991	0.11868	1.29534	0.996	

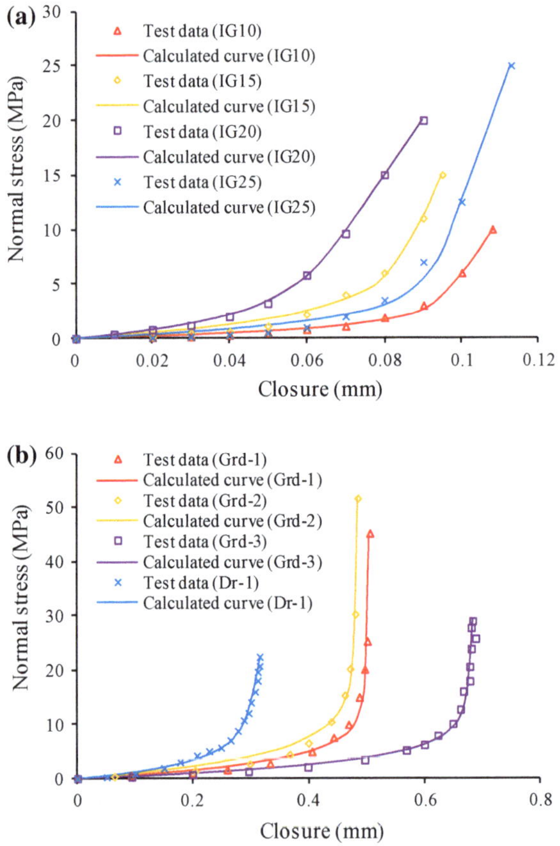

Fig. 3.23 Comparisons between measured data [from **a** Matsuki et al. (2001) and **b** Malama and Kulatilake (2003)] and results calculated from Eq. 3.80 (Liu et al. 2013). Items in *parenthesis* refer to the sample numbers from Table 3.8

areas as stress changes. The upper limit of $K_{e,f}$ should be of the same order of magnitude as that for the rock matrix.

As a relatively simple model, the TPHM has two unique aspects in describing rock deformations. Firstly, it was developed by extending the stress-strain relationship described by Hooke's Law; the TPHM is reduced to the conventional form of Hooke's Law when the soft-part portion is zero. Thus, it can be used to derive relevant constitutive relationships for hydraulic and mechanical properties using a consistent set of parameters with clear physical meanings (Liu et al. 2009; Zhao and Liu 2012), because the corresponding stress-strain relationship is the foundation for deriving other constitutive relationships. Secondly, comparisons between results calculated from the TPHM and test data show that the TPHM is quite general and can be applied to both fractures and porous media.

3.3.1.2 Two-Phase Flow Properties

Multiphase flow occurs in a number of practical applications associated with fractured rock, such as geological sequestration of CO_2 and nuclear waste disposal. However, most of the previous studies on the stress dependence of fracture properties have focused on single-phase flow properties, such as permeability (and fracture aperture or closure). A systematic study on relationships between two-phase or multiphase flow properties for fractures (such as relative permeabilities for different phases) and normal stress was recently reported by Liu et al. (2013).

Even without considering the stress-dependence, determination of two phase flow properties is a challenging task for fractures. There are currently two kinds of approaches to developing the related constitutive relationships (between capillary pressure, relative permeability and saturation) for two phase flow. One is a porous-medium approach, in which constitutive relationships developed for porous media are simply borrowed or modified for either single fractures or fracture networks (e.g., Therrien and Sudicky 1996). However, the physical meaning of relevant parameters in these relationships is not always clear for fractures, because the geometry (and connectivity) of fracture apertures is essentially two dimensional on the fracture plane, whereas that of pores in porous media is three dimensional. The second approach, called the pipe-flow approach, has been developed based on similarities between the observed multiphase-flow behavior within fractures and pipes, including the fact that both wetting and non-wetting phases are discontinuous and yet mobile (Wong et al. 2008; Weerakone et al. 2011), or stratified owing to density differences within horizontally-inclined fractures (Fourar and Lenormand 1998; Indraratna et al. 2002). This approach may be valid when fluid-phase distributions are not strongly dependent on fracture apertures (or capillary force). From laboratory test results for nitrogen/water flow within horizontal fractures, Diomampo (2001), however, did not observe pipe-flow behavior. It is obvious that different regimes may exist for two phase flow within fractures. The focus here is on two-phase flow within a horizontal fracture when flow paths are mainly determined by capillarity (or the aperture distribution). This subsection will present the TPHM-based closed-form relationships between two-phase flow properties and normal stress for horizontal fractures that were proposed by Liu et al. (2013).

When flow paths in a horizontal fracture are mainly determined by capillarity, the concept used to derive two-phase flow properties for porous media can be adopted here, although the related relationships for porous media cannot be directly borrowed, as previously indicated. The standard approach used in porous media is based on the local equilibrium assumption that capillary pressure is uniformly distributed within the pore space corresponding to representative elementary volume (e.g., Mualem 1976). The assumption essentially implies that liquid distribution for a given phase is completely controlled by capillary force and is independent of liquid flux. That is why capillary pressure and relative permeability can be expressed as functions of saturation only, as discussed in Chap. 2.

However, in many cases, the local equilibrium assumption does not hold, especially when instability (or fingering) occurs. For example, for the upward flow of CO_2 in a saline aquifer at a geological CO_2 sequestration site, significant CO_2 fingering will occur because CO_2 is lighter than ambient brine. When a numerical grid block contains a number of CO_2 fingers (which is generally the case, because the grid-block size cannot be too small for practical applications), the local equilibrium assumption will be violated at the grid-block scale. Chapter 2 proposes a new theory for multiphase flow to deal with this challenging issue, based on the optimality principle that unstable liquid-flow patterns are self-organized in such a way that total flow resistance is minimal. In this case, the relative permeability is a function of not only saturation, but also liquid flux. Nevertheless, it is appropriate to apply the local equilibrium assumption to individual horizontal fractures that are not subject to gravitational instability for flow process.

The capillary pressure P_c, defined as the difference between non-wetting-phase and wetting-phase pressures, is given by the Young-Laplace equation for a fracture element (with aperture B) at the fluid interface:

$$P_c = \frac{T_s \cos\beta'}{2B} \tag{3.82}$$

where T_s is surface tension and β' is contact angle. Based on the local equilibrium assumption and above equation, fracture apertures larger than $T_s \cos\beta'/(2P_c)$, for a given capillary pressure P_c, are generally occupied by non-wetting phase and the rest by wetting phase. This reasoning is the foundation for deriving the capillary pressure-saturation relationship, as discussed in Chap. 2. Note that B is fracture aperture at a given location of the fluid interface within a fracture plane and b (often used in Sect. 3.3.1.1) is the average aperture for a fracture. Their relation is mathematically given in Eq. 3.88 to be discussed later.

The relationship between capillary pressure and fluid saturation is determined by the probability density function (PDF) of the fracture aperture (within a single fracture) and its dependence on stress. Approximately Gaussian or truncated-Gaussian aperture distributions are commonly observed in both natural and man-made fractures (Walsh et al. 2008). As shown in Fig. 3.24, with increasing stress, small apertures are reduced to zero and the PDF is close to a truncated-Gaussian distribution. A number of researchers reported that the PDF under normal load might be described by log-normal distributions to characterize its skewed behavior (e.g., Pruess and Tsang 1990). Recently, Sharifzadeh et al. (2008) developed a new method for measuring fracture-aperture distributions and reported measurement results for several artificial fractures. They found that the observed PDFs are similar to Poisson or log-normal distributions. Thus, different kinds of distribution may be used for describing the PDF for fracture apertures. In the study of Liu et al. (2013), for simplicity, it was assumed that the fracture-aperture PDF is described by a truncated-Gaussian distribution:

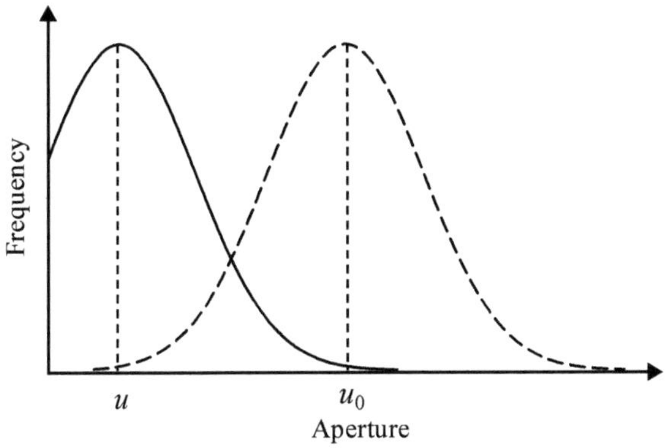

Fig. 3.24 Gaussian (*dashed curve*) and truncated-Gaussian (*solid curve*) aperture distributions (Liu et al. 2013). The *solid curve* is subject to a larger normal stress. The parameters u_0 and u are means for the Gaussian distribution for the *dashed curve* and the corresponding Gaussian distribution for the *solid curve*, respectively. They correspond to the largest probability values

$$f(B) = \frac{1}{\sqrt{2\pi}\delta} e^{-\frac{(B-u)^2}{2\delta^2}} \quad \text{for } B \geq 0 \tag{3.83}$$

where u and δ are mean and standard deviation, respectively, for the corresponding non-truncated Gaussian distributions. Note that the integration of $f(B)$ within the valid range of $B \geq 0$ is less than one, as a result of truncation. A mathematically rigorous PDF requires the integration to be one. To meet this requirement, the right-hand side of Eq. 3.83 can be multiplied by a normalized factor. However, as demonstrated by derivation procedures to be discussed later, the direct use of Eq. 3.83 is valid for this study, because normalization processes are always involved when this equation is used.

Based on the local equilibrium assumption and Eq. 3.83, the non-wetting phase saturation (S_{nw}) for a capillary pressure given by Eq. 3.82 is obtained as (Liu et al. 2013):

$$S_{nw} = \frac{\int_B^\infty xf(x)dx}{\int_0^\infty xf(x)dx} = \frac{\frac{\delta}{\sqrt{2\pi}} e^{-t_b^2} + \frac{u}{2} erfc(t_b)}{\frac{\delta}{\sqrt{2\pi}} e^{-t_{00}^2} + \frac{u}{2} erfc(t_{00})} \tag{3.84}$$

$$t_b = \frac{B-u}{\sqrt{2}\delta}; \quad t_{00} = \frac{-u}{\sqrt{2}\delta} \tag{3.85}$$

where $f(x)$ is the PDF defined by Eq. 3.83 and x here is a dummy variable.

The corresponding wetting-phase saturation (S_w), by definition, for a two-phase flow system is:

$$S_w = 1 - S_{nw} \tag{3.86}$$

In Eq. 3.84, fracture aperture B can be replaced by capillary pressure P_c using Eq. 3.82. If a capillary pressure $P_c = P^*$ is known for a fracture aperture $B = B^*$ that corresponds to fracture elements at the interface between wetting and non-wetting phases, then Eq. 3.82 can be rewritten as:

$$\frac{P_c}{P^*} = \frac{B^*}{B} \tag{3.87}$$

Equations 3.82 (or 3.87), 3.84–3.86 together comprise the relationship between P_c, S_{nw}, and S_w. For a given capillary pressure, Eq. 3.82 or Eq. 3.87 is used to estimate fracture aperture B associated with the interface between wetting and non-wetting phases. Then, saturations are calculated from Eqs. 3.84 to 3.86.

To determine the stress dependence of capillary pressure–saturation relationships, it must be known how the fracture-aperture PDF changes with normal stress. The latter corresponds to complex mechanical deformations because of the variability in fracture asperity and contact areas. Recently, Walsh et al. (2008) conducted a detailed numerical simulation of mechanical deformation within a fracture under different normal stresses. They found that the aperture PDF becomes more skewed with increasing normal stress, but is still closer to a truncated-Gaussian distribution than a log-normal distribution. They also noticed that the PDF evolution can be approximately described by a so-called penetration model, in which fracture aperture changes uniformly while aperture values are limited to being zero or positive. The same model was used by other researchers in determining stress-dependent fracture properties (e.g., Brown 1987; Oron and Berkowitz 1998). Thus, the penetration model is adopted here. In other words, parameter δ is treated as being independent of stress, but u will change with stress, as illustrated in Fig. 3.24, where the solid curve represents a PDF corresponding to a larger stress. The parameter u can be related to average aperture b (Eq. 3.79) by (Liu et al. 2013):

$$b = \frac{\int_0^\infty Bf(B)dB}{\int_0^\infty f(B)dB} = u + \frac{\sqrt{\frac{2}{\pi}}\delta e^{-t_{00}^2}}{erfc(t_{00})} \tag{3.88}$$

where t_{00} is defined in Eq. 3.85. Note that in Eq. 3.88, $Bf(B)dB$ represents the volume occupied by fracture segments with apertures from B to B + dB. The stress dependence of b is given in Eq. 3.79. For a given value of b, u can be derived from Eq. 3.88, but an iterative numerical procedure is needed. Then the capillary pressure-saturation relationship can be calculated through the dependence of u on stress. Figure 3.25 shows curves of capillary pressure (as a function of wetting-phase saturation) for two different stresses (or u values). As expected, the capillary pressure curve strongly depends on the normal stress.

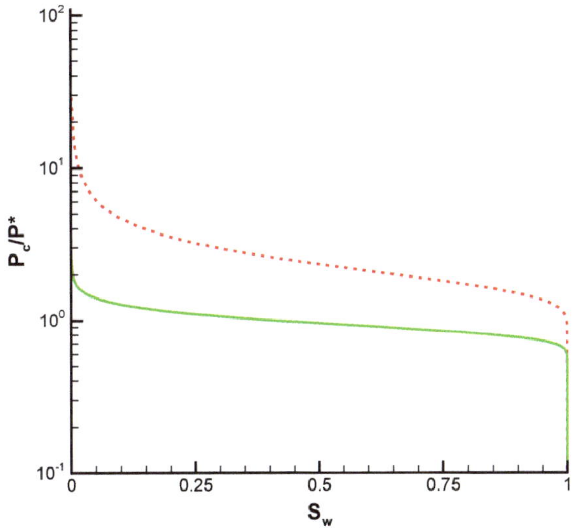

Fig. 3.25 Normalized capillary pressure (P_c/P^*) as a function of wetting-phase saturation (S_w) for two stresses, with the *solid curve* corresponding to a smaller normal stress (Liu et al. 2013). The parameter values are: $u = 0.145$ mm (for the *solid curve*) and 0.045 mm (for the *dashed curve*), and $\delta = 0.03$ mm (parameters as defined in the text)

The relationship between fracture permeability and stress is given in Eqs. 3.79 and 3.81. Modeling of two-phase flow processes also requires the relative permeability for different phases. As previously indicated, studies on fracture relative permeability are very limited and no systematic study has been found in the literature, except Liu et al. (2013), that investigates stress-dependent fracture relative permeability. The relationships of Liu et al. (2013) will be presented between fracture relative permeability, saturation, and stress for a two-phase flow system.

The general approach to developing closed-form relationships in porous media involves two key aspects, as discussed in details in Chap. 2. Firstly, based on the PDF of pore-size distribution, a conceptual model of two-phase flow through the medium is developed. There are currently two commonly-used conceptual models. One was proposed by Mualem (1976), in which the spatial distribution of soil pores is completely random. The other one is the model developed by Burdine (1953), in which a perfect spatial correlation of soil pores is assumed, such that flow paths can be conceptualized as a group of parallel capillary tubes. In both models, pore size is determined from capillary pressure–saturation relationship using the Young-Laplace equation that suggests that pore-size is inversely proportional to the capillary pressure. Note that pore size for porous media corresponds to fracture aperture herein. The second aspect of this general approach is to make further corrections to the results, determined from these conceptual models, by multiplying them with a tortuosity factor. The rationale behind this is that in those conceptual models, flow paths are assumed to be straight, while in reality, they are tortuous as a result of the geometric complexity of the pore spaces. The tortuosity factor is generally assumed to be a power function of saturation for a given phase and the exponent value needs to be empirically determined from measurements. In the current study, this general approach is followed to develop relative permeability for a horizontal fracture.

Note that the corresponding relationships cannot be directly borrowed from porous media, simply because geometry (and connectivity) of fracture apertures is essentially two dimensional and is different from that of pores in porous media, which is three dimensional.

Burdine's (1953) model was employed herein, based on the consideration that flow paths in fractures are close to parallel capillary tubes that have been implied by commonly-reported long-range correlations (fractal behavior) of aperture distribution and observed channelized multiphase flow patterns (e.g., Walsh et al. 2008; Brown 1987; Power and Tullis 1992; Chen and Horne 2006). The wetting phase relative permeability, k_{wr}, is expressed as (Burdine 1953):

$$k_{wr} = \frac{\int_{B_{\min}}^{B} \frac{1}{P_c^2} dV}{\int_{B_{\min}}^{\infty} \frac{1}{P_c^2} dV} S_w^{m'} \tag{3.89}$$

where the power function term for S_w is the tortuosity factor with m' as an empirical parameter, and dV is the volume element corresponding to fracture aperture B associated with capillary pressure P_c:

$$dV \propto Bf(B)dB \tag{3.90}$$

In Eq. 3.89, B_{min} refers to a critical fracture aperture. When the wetting phase is confined to fracture apertures smaller than B_{min}, the wetting phase becomes discontinuous and thus immobile. Inserting Eqs. 3.82 and 3.90 into Eq. 3.89 yields:

$$k_{wr} = S_w^{m'} \frac{\int_{B_{\min}}^{B} B^3 f(B)dB}{\int_{B_{\min}}^{\infty} B^3 f(B)dB} \tag{3.91}$$

Similarly, the relative permeability for the non-wetting phase (k_{nwr}) can be written as:

$$k_{nwr} = S_{nw}^{m'} \frac{\int_{B}^{B_{\max}} B^3 f(B)dB}{\int_{0}^{B_{\max}} B^3 f(B)dB} \tag{3.92}$$

where B_{max} is the other critical fracture aperture. When the non-wetting phase is confined to fracture apertures larger than B_{max}, the non-wetting phase becomes immobile as a result of blockage by the wetting phase. Note that B_{min} and B_{max} correspond to residual saturations for wetting and non-wetting phases, respectively.

As previously discussed, the fracture-aperture PDF can be presented by a truncated-Gaussian distribution given in Eq. 3.83. Before presenting the relative permeability relationships subject to the PDF, it is useful to define several derived integration functions:

$$I_1(x,y) = \int_x^y e^{-t^2} dt = \frac{\sqrt{\pi}}{2}[erfc(x) - erfc(y)]$$

$$I_2(x,y) = \int_x^y te^{-t^2} dt = \frac{1}{2}\left[e^{-x^2} - e^{-y^2}\right]$$

$$I_3(x,y) = \int_x^y t^2 e^{-t^2} dt = \frac{1}{2} I_1(x,y) + \frac{1}{2}\left(xe^{-x^2} - ye^{-y^2}\right)$$

$$I_4(x,y) = \int_x^y t^3 e^{-t^2} dt = \frac{1}{2}\left\{e^{-x^2}\left(x^2+1\right) - e^{-y^2}\left(y^2+1\right)\right\}$$

$$I_t(x,y) = \int_x^y t^3 f(t) dt = \frac{u^3}{\sqrt{\pi}} I_1(x,y) + 3\delta u^2 \sqrt{\frac{2}{\pi}} I_2(x,y) + \frac{6\delta^2 u}{\sqrt{\pi}} I_3(x,y) + 2\delta^3 \sqrt{\frac{2}{\pi}} I_4(x,y) \tag{3.93}$$

where u and δ are parameters associated with the fracture-aperture PDF defined in Eq. 3.93. Also in this equation, x, y, and t are dummy variables.

Inserting Eq. 3.83 into Eqs. 3.91 and 3.92 and using Eq. 3.93, the relative-permeability expressions are obtained for wetting and non-wetting phases as (Liu et al. 2013):

$$k_{wr} = S_w^{m\prime} \frac{I_t(t_{\min}, t_b)}{I_t(t_{\min}, \infty)} \tag{3.94}$$

$$k_{nwr} = S_{nw}^{m\prime} \frac{I_t(t_b, t_{\max})}{I_t(t_{00}, t_{\max})} \tag{3.95}$$

$$t_{\min} = \frac{B_{\min} - u}{\sqrt{2}\delta}; \quad t_{\max} = \frac{B_{\max} - u}{\sqrt{2}\delta} \tag{3.96}$$

where t_b is defined in Eq. 3.85.

Relative permeability (Eqs. 3.94 and 3.95) is a function of the parameter u that is related to stress through fracture aperture (Eq. 3.88). The relationship between average fracture aperture and stress is described by the TPHM (Eq. 3.79). Relative permeability is also a function of B_{min} or B_{max}. For the non-wetting phase, the residual saturation can be determined using the following percolation procedure. Consider a horizontal fracture to be initially filled with a wetting phase. Then the largest apertures occupied by the wetting phase are replaced by a non-wetting phase step by step until a continuous non-wetting flow path is formed. The corresponding non-wetting phase saturation is the residual saturation. Since fracture aperture changes uniformly (Fig. 3.24), regions within a fracture that are occupied by a residual non-wetting phase remain the same under different stress conditions. In other words, B_{max} follows:

$$B_{\max} - B_{\max,ref} = u - u_{ref} \tag{3.97}$$

where the subscript *ref* refers to a reference condition under which related parameter values can be estimated from testing data. Applying a similar consideration to the wetting phase, that the smallest apertures initially occupied by the non-wetting phase are replaced by a wetting phase step by step until a continuous wetting flow path is formed, gives:

$$B_{\min} - B_{\min,ref} = u - u_{ref} \tag{3.98}$$

where the smallest value for B_{min} is limited to being zero. Equation 3.98 is based on an argument that fracture regions containing a residual wetting phase (including fracture space that is closed as result of increasing stress) remain unchanged with changing stress. Unlike the non-wetting phase, the situation for the wetting phase is more complex, because the residual regions involve opening and closing fracture apertures with changing stress, which can considerably alter wetting-phase connectivity in some cases. Thus, Eq. 3.98 is considered to be a first-order approximation at this point. It is also useful to note that $B_{min} = 0$ and $B_{max} = \infty$ correspond to zero residual saturations for wetting and non-wetting phases, respectively. As an example, Fig. 3.26 shows relative permeability curves for two stresses (or different u values), indicating considerable sensitivity to the stress change.

So far the closed-form relationships between capillary pressure, saturation, and stress have been developed for two-phase flow in a horizontal fracture. The dependence of these relationships on stress is realized through parameter u that

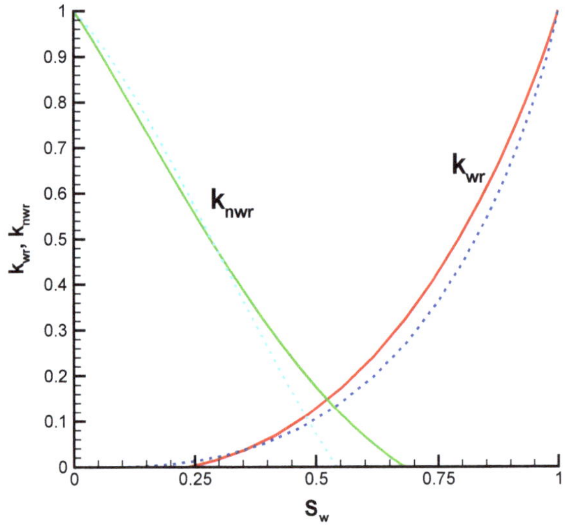

Fig. 3.26 Relationships between relative permeability and saturation for two different normal stresses with the *dashed curve* corresponding to a larger stress (Liu et al. 2013). The parameter values are: u = 0.145 mm (for the *solid curve*) and 0.045 mm (for the dashed curve), B_{min} = 0.13 mm (for the *solid curve*) and 0.03 mm (for the *dashed curve*), B_{max} = 0.165 mm (for the *solid curve*) and 0.065 mm (for the *dashed curve*), and δ = 0.03 mm (parameters as defined in the text)

changes with stress. Unfortunately, testing data on the stress dependence of multiphase flow properties for fractures are very rare, if they even exist, in the literature. Liu et al. (2013) therefore evaluated these relationships for fractures that are not subject to mechanical deformation. Two aspects in this evaluation are of particular interest: (1) whether the relationships can satisfactorily represent experimental observations; and (2) whether a tortuosity factor with the same m' value (Eqs. 3.94 and 3.95) can be applied to different phases for different tests. As indicated in Chap. 2, the universal m' value has been used in porous media (Brooks and Corey 1964).

The data set of Chen and Horne (2006) is employed here for evaluating the theoretical results, because this data set, including information on the fracture-aperture PDF (Gaussian distribution), and the corresponding experimental procedures are very well documented (Liu et al. 2013). Chen and Horne (2006) conducted concurrent, steady-state, air-water flow experiments in horizontal fractures at the room temperature. The fractures for the experiments, with dimensions of around 35 cm × 15 cm, were created by mounting a rough glass plate on top of a smooth aluminum plate, confined by a metal frame bolted to the bottom plate. All measurements were digitized automatically with a high-speed data acquisition system. Two fractures with distinct surface roughness, i.e., one is homogeneously rough (HR) and the other randomly rough (RR), were used in the tests; the former does not have correlation in spatial-aperture distribution and the latter does. Their test results show strong channelized flow behavior and they also reported relative permeability data as a function of water saturation for both the wetting phase (water) and the non-wetting phase (air). Details of the data set and its collection procedure can be found in Chen and Horne (2006). Data for a smooth fracture were also reported in Chen and Horne (2006), but are not used in this study because all natural fractures are rough (Liu et al. 2013).

Figure 3.27 shows comparisons between the relative permeability data of Chen and Horne (2006) and results calculated from Eqs. 3.94 and 3.95 for two (RR and HR) fractures. Parameter values for u and δ are taken directly from Chen and Horne (2006). Values for parameters B_{min}, B_{max}, and m' in Eqs. 3.94–3.96 are adjusted to achieve curve fitting. In general, matches between calculated results and test data are reasonable, suggesting that the theoretical relationships can satisfactorily represent the test data. It is especially of interest to note that a single m' value of 1.0 was used for both wetting and non-wetting phases for the two fractures. Note that $m' = 2$ in the widely-used Brooks and Corey (1964) model that was developed for soils from Burdine's (1953) model, which means that for a given saturation soils have smaller tortuosity-factor values than fractures. This makes sense physically, because flow paths are more tortuous in three-dimensional pore space for soils than in essentially two-dimensional fracture space. However, it is acknowledged that further evaluation of the relationships with more data sets, when they become available, is necessary to further confirm the result of $m' = 1$ for fractures (Liu et al. 2013).

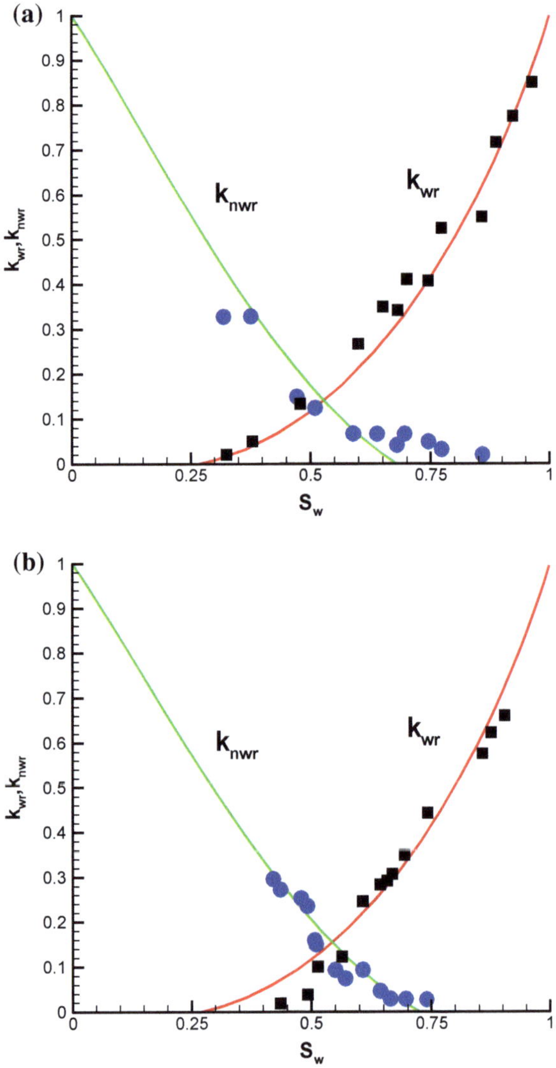

Fig. 3.27 Comparisons between the relative permeability data reported in Chen and Horne (2006) (*black squares* and *blue circles*) and calculated results (*solid curves*) from Eqs. 3.94 and 3.95 for two fractures (Liu et al. 2013): **a** Homogeneously rough; and **b** Randomly rough. Parameter values are: **a** u = 0.145 mm, B_{min} = 0.13 mm, B_{max} = 0.165 mm, and δ = 0.03 mm; **b** u = 0.24 mm, B_{min} = 0.22 mm, B_{max} = 0.28 mm, and δ = 0.05 mm (parameters as defined in the text)

3.3.2 Internal Swelling Stress

Fracture deformation becomes more complex for some subsurface media subject to swelling or shrinkage. For example, carbon dioxide (CO_2) injected into deep, unminable coal seams can give rise to enhanced production of coal bed methane by displacing methane (CH_4) from coal, but at the same time causes the swelling of coal as well. One key parameter for CO_2 injection is coal permeability, because high coal permeability is required for sufficient and practical injectivity of CO_2 into coal seams and for efficient recovery of CH_4 (Cui et al. 2007). Note that coal is a fractured medium and its permeability mainly comes from fractures (cleats). Results from field and laboratory experiments indicate that coal permeability can change significantly as controlled by at least two distinct mechanisms (Harpalani and Zhao 1989; Palmer and Mansoori 1998; Mavor and Vaughn 1997; Robertson 2005; White et al. 2005; Lin et al. 2008). The first one is the gas pressure increase, which tends to mechanically open coal cleats (fractures) and thus enhance coal permeability. The second mechanism is the adsorption of CO_2 into coals, which induces swelling in the coal matrix and thus reduces coal permeability by narrowing and even closing fracture (cleat) apertures.

The other example is shale rock that swells (or shrinks) with increasing (or decreasing) water content. Shale rock has been considered worldwide as a rock type for geological disposal of high-level nuclear waste. Recovery of natural gas from a shale gas reservoir with hydraulic fracturing also involves water-based fracturing fluid flowing from hydraulic fractures to surrounding shale rock matrix. Thus, rock swelling has important implications for these two applications related to shale formations.

It is obvious that relationships between fracture deformation (and fracture properties) and rock matrix swelling are important for understanding and modeling coupled hydro-mechanical processes in all the examples mentioned above. In this subsection, we will present such a relationship for a rock involving swelling that is based on the TPHM and the concept of internal swelling stress (Liu and Rutqvist 2010; Liu et al. 2011).

3.3.2.1 Effective Stress for Fractures Involving Rock Swelling

It is relatively straightforward to deal with the effective stress for rock matrix involving swelling. The treatment is essentially the same as handling volume expansions of the rock matrix owing to temperature increases (e.g., Jaeger et al. 2007). However, some special consideration needs to be given to fractures when swelling processes are involved. The content of this subsection is based on the study of Liu and Rutqvist (2010).

Based on Biot's theory, the effective stress is defined as (Jaeger et al. 2007)

$$\sigma = \sigma_T - \alpha_B P \tag{3.99}$$

where σ_T is total stress, P is fluid pressure, and α_B is Biot's coefficient of effective stress. Note that compressive stress is again considered positive here.

The traditional concept of effective stress (Eq. 3.99) cannot effectively describe the often observed changes in fracture permeability (aperture) with rock-matrix swelling. For example, let's consider a rock containing an infinite fracture (generally assumed for modeling flow and transport in fractured rock). In this case, matrix swelling, based on Eq. 3.99, will not affect the effective stress under conditions of the constant confining (total) stress and pore pressure that are commonly

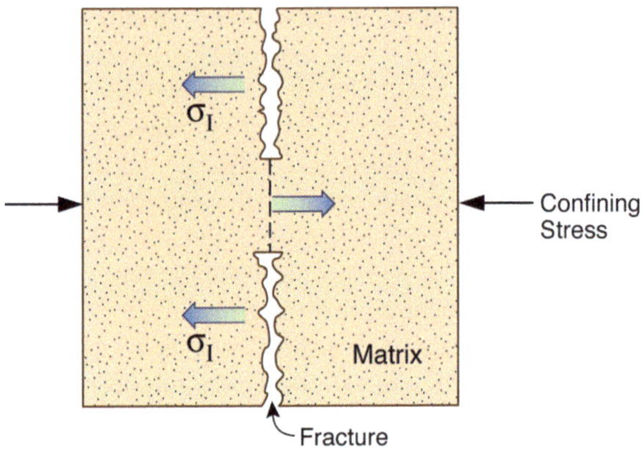

Fig. 3.28 A schematic description of internal swelling stress (Liu and Rutqvist 2010). The *arrows* σ_I correspond to stresses imposed on the left part of the rock as a result of swelling

used in laboratory measurements. In other words, one cannot relate matrix swelling and the resultant fracture-permeability changes to the effective stress defined in Eq. 3.99.

In general, rock matrix blocks are not completely separated from each other by fractures. Figure 3.28 shows a simplified horizontal cross section of a rock (involving swelling) with two adjacent vertical fractures, separated by a matrix "bridge" that connects matrix blocks on the different sides of fractures. During matrix swelling, fractures are compressed, because they are weak and soft structures within the rock, and therefore an additional force (corresponding to stress σ_I) will be imposed on the fractures. At the same time, the matrix bridge is subject to an additional force in the opposite direction to σ_I. If these two forces are completely balanced, fractures will be subject to this additional stress σ_I, while confining stress remains unchanged. Because this stress largely results from internal structures (or the connectivity of matrix blocks) within the rock body and can be internally balanced under constant confining stress conditions, it is called "internal swelling stress". In this case, the effective stress for fractures should be given as (Liu and Rutqvist 2010)

$$\sigma = \sigma_T - \alpha_B P + \sigma_I \tag{3.100}$$

Note that σ_I is positive for matrix swelling and negative for matrix shrinkage. The concept of "internal swelling stress" was first put forward by Liu and Rutqvist (2010) who derived a similar effective stress equation for coal seams associated with swelling.

The concept of internal swelling stress implies that rock-matrix strain resulting from swelling (ε_s) can be divided into two parts:

$$\varepsilon_s = \varepsilon_{sB} + \varepsilon_{sI} \tag{3.101}$$

where ε_{sI} is the strain corresponding to the internal swelling stress, and ε_{sB} is the strain contributing to the bulk strain in the fractured rock (involving swelling) generally measured in the laboratory. It is ε_{sI} (a portion of ε_s) that results in fracture permeability (or aperture) changes under constant confining stress conditions.

The relationship between ε_s and ε_{sI} may be a complex function of matrix block connectivity within the rock and other relevant factors. As a first approximation, Liu and Rutqvist (2010) assumed the ratio between the two strains to be a constant:

$$\varepsilon_{sI} \approx f\varepsilon_s \tag{3.102-1}$$

where f is a value between zero and one and can be determined from measurements. This treatment will be evaluated against laboratory test results and more studies may be needed to develop more rigorous relationships between f and other properties in the future. Based on Hooke's Law, σ_I can be related to swelling strain by

$$\sigma_I = f\varepsilon_s K_m \tag{3.102-2}$$

where K_m is the bulk modulus for rock matrix and can be stress dependent.

3.3.2.2 A Data Analysis

This subsection demonstrates the usefulness of the internal swelling stress concept by comparing the theoretical results with selected laboratory measurements for a shale (or clay) rock. The content of this subsection is based on the study of Liu et al. (2011).

Davy et al. (2007) reported on laboratory measurements of single fractures within macro-cracked Callovo-Oxfordian argillite samples subject to both confinement and water-induced swelling. The data set provides a unique opportunity to examine the proposed formulations for estimating fracture permeability as a function of effective stress that considers the effects of swelling.

For water permeability tests, the experimental procedure was designed as follows (Davy et al. 2007). Apply an initial continuous fluid flow through the fracture, and then superimpose an additional pulse flow for permeability measurements made at each confining pressure level, either right after loading or after several hours at a given confining pressure (the total stress), or right after unloading. For all of the tests, a posteriori visual inspection of the water-permeability samples shows limited water penetration in the argillite sample bulk. Figure 3.29 shows the test procedure in terms of the changes in confining pressure and fracture closure for Sample 2 (Davy et al. 2007). Although water permeability measurements are provided for two samples (samples 2 and 5) in Davy et al. (2007), test results for Sample 2 are analyzed only in Liu et al. (2011), because Sample 2 is subject to a more complex test procedure (Fig. 3.38). For a given confining pressure, the fracture closure increases from Point 1 to 2, which cannot be explained based on elastic deformation and is very likely due to water-induced plastic deformation at the beginning of the test. Therefore, the analysis focuses on data points after Point 2. Liu et al. (2011) also assumed elastic deformation in that data range—mainly justified by the fact

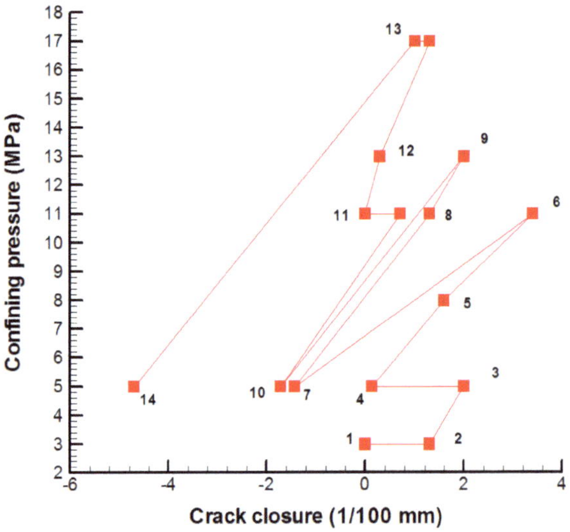

Fig. 3.29 Test procedure of water permeability measurement (in terms of confining pressure and fracture (crack) closure) for sample 2 of Davy et al. (2007) (Liu et al. 2011). The *numbers* in the figure indicate the chronological order of points

that our analysis based on the elastic deformation seems to be able to explain the majority of experimental observations.

In Fig. 3.29, Points 3, 4, 7, 10 and 14 correspond to the same confining pressure but different amounts of swelling (measured as the difference in crack closure between a given point and Point 3). We believe that it is largely due to the transient behavior of water flow from fractures into the rock matrix. A longer time corresponds to a larger water penetration depth into the rock matrix near the fracture, and therefore to a larger rock volume involving swelling. Note that during the water permeability measurement, water was injected into the fracture. For simplicity, Liu et al. (2011) assumed that water penetration depth as a function of time can be estimated from Eq. 1.150 and Darcy's law is approximately valid here. Under ponding conditions on the ground surface, Eq. 1.150 indicates that the cumulative amount of water infiltrating into unsaturated soil with uniform initial water saturation is proportional to the square root of time. Consequently, if we view the fracture wall as the ground surface, then approximate the water penetration depth by the amount of accumulative infiltrating water (in depth) divided by the difference between saturated and initial water contents, and further assume that swelling within the water-penetrating zone is uniform and occurs simultaneously once water content is increased, then the total swelling, SW, will be proportional to water penetration depth, or

$$SW = At^{1/2} \tag{3.103}$$

where A is a constant and t is time. The above equation (with A = 3.08E−2 mm/d$^{-1/2}$) seems to fit observed swelling for Points 3, 4, 7, 10, and 14 (corresponding to different times) satisfactorily (Fig. 3.30), indicating that the above reasoning is valid. Note that the observed crack-closure value in Davy et al. (2007) is a combination of rock

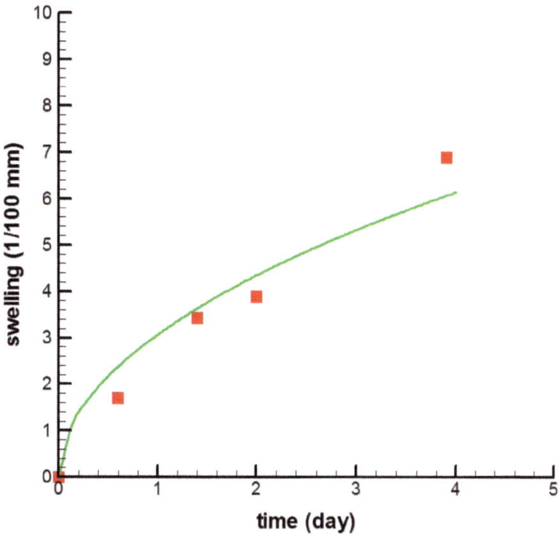

Fig. 3.30 Comparison of observed swellings for *points* (3, 4, 7, 10, and 14) with those calculated from Eq. 3.103 (*solid curve*) (Liu et al. 2011)

swelling and the corresponding change in fracture aperture. However, as a result of the low water permeability of fracture, the fracture aperture value (estimated from the cubic law) is negligibly small, only on the order of 1E−3 mm. Therefore, the swelling is approximated here by the observed crack closure.

Because fracture permeability is proportional to the cube of fracture aperture (Eq. 3.81), the fracture permeability k, based on the combination of Eqs. 3.79 and 3.81, is given by (Liu et al. 2011)

$$\left(\frac{k}{k_0}\right)^{1/3} = \frac{b_{0,e}}{b_0}\left(1 - \frac{\sigma_n}{K_{e,f}}\right) + \frac{b_{0,t}}{b_0}\exp\left(-\frac{\sigma_n}{K_{t,f}}\right) \tag{3.104}$$

where k_0 is the permeability corresponding to b_0. Assuming that the entire fracture aperture is "soft", one has

$$\left(\frac{k}{k_0}\right)^{1/3} = \exp\left(-\frac{\sigma_n}{K_{t,f}}\right) \tag{3.105}$$

Given the fact that shale or clay rock is generally viewed as soft rock, it seems logical to use Eq 3.105 for fractures in clay rock. This is supported by a number of laboratory measurements that show linear relationships between the log of measured fracture permeability and stress (e.g., Blumling et al. 2007; Zhang and Rothfuchs 2008; Popp et al. 2008). However, data reported by Jobmann et al. (2010) seem to indicate that permeability relationships are better represented by a curve with additional contribution from the hard part. For simplicity, we focus on Eq. 3.105, which seems to be reasonable for most clay (shale) studies reported in the literature, while Eq. 3.104 may be employed for more general cases (Liu et al. 2011).

When confining and pore pressures are constant, fracture permeability purely due to swelling may be obtained from Eqs. 3.100 and 3.105 and given as

$$\left(\frac{k}{k_3}\right)^{1/3} = \exp\left(-\frac{\Delta\sigma_I}{K_{t,f}}\right) \tag{3.106}$$

where k_3 is the permeability at Point 3 and $\Delta\sigma_I$ is the difference in internal swelling pressure between a given point and Point 3. Using the definition of the internal swelling stress (Eqs. 3.102-1 and 3.102-2) together with Eq. 3.103, the difference in internal swelling pressure is given as

$$\Delta\sigma_I = f\varepsilon_s K_m = fK_m\frac{SW}{L} = \left(\frac{fK_mA}{LK_{t,f}}\right)K_{t,f}t^{1/2} = BK_{t,f}t^{1/2} \tag{3.107}$$

where L is fracture spacing (approximated by the rock-sample's radius in Davy et al. (2007)), and B is a constant. Combining Eqs. 3.106 and 3.107 yields (Liu et al. 2011)

$$\left(\frac{k}{k_3}\right)^{1/3} = \exp\left(-Bt^{1/2}\right) \tag{3.108}$$

Again, Eq. 3.108 fits fairly well the observations for the data points 3, 4, 7, 10 and 14 (in Fig. 3.29) that correspond to the same confining pressure, but different amounts of swelling (Fig. 3.31).

Equation 3.108 is applicable only when confining pressure (or total stress) and the pore pressure of water in the fracture are constant. Pore pressure changes are negligible here in comparison with the much larger confining pressure in the water

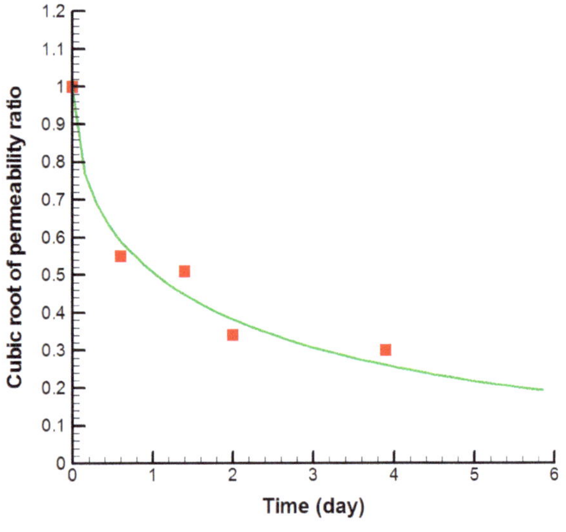

Fig. 3.31 Match between observed values for $(k/k_3)^{1/3}$ for *points* (3, 4, 7, 10 and 14) with those calculated from Eq. 3.108 (*solid curve*) (Liu et al. 2011)

permeability experiments of Davy et al. (2007). This can be justified by the observation that fracture permeability changes are mainly determined by confining pressure and swelling (Davy et al. 2007).

A more general permeability relationship (that considers the effects of both confining pressure and swelling) can be obtained by combining Eqs. 3.105, 3.100 and 3.108 (Liu et al. 2011):

$$\left(\frac{k}{k_3}\right)^{1/3} = \exp\left(-\frac{\Delta\sigma_t}{K_{t,f}} - Bt^{1/2}\right) \tag{3.109}$$

The only unknown in the above equation is $K_{t,f}$ which can be estimated from the permeability data as a function of both confining pressure and time (Davy et al. 2007). We are especially interested in whether Eq. 3.109 is sufficient to represent the data. The estimated (or fitted) $K_{t,f}$ is 16 MPa that is generally consistent with the range of estimated $K_{t,f}$ in Table 3.8. Figure 3.32 shows a comparison between measured and estimated permeability values as a function of time. Note that for a given time in Fig. 3.32, there are two data points corresponding to the observed and estimated values, respectively. Given the complexity of the experimental processes, the agreement is remarkable, supporting the validity of the relevant constitutive relationships.

To further examine the usefulness of our generalized effective stress (Eq. 3.100), Fig. 3.33 shows k as a function of effective stress difference (between a given point and Point 3) calculated by

$$\Delta\sigma = \Delta\sigma_T + \Delta\sigma_I = \Delta\sigma_T + K_{t,f}Bt^{1/2} \tag{3.110}$$

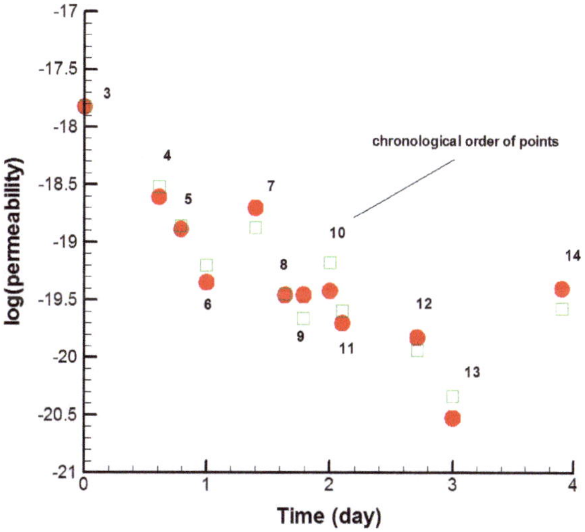

Fig. 3.32 Comparisons between observed and simulated fracture permeability changes as a function of time. The *solid circles* are measurements (Liu et al. 2011). The numbers in the figure have the same meaning as in Fig. 3.29 and each number here corresponds to a pair of measured and simulated values for a given time

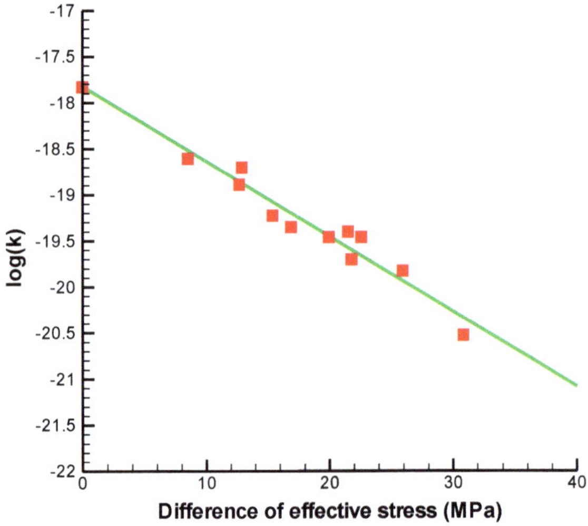

Fig. 3.33 Comparisons between observed and simulated relationships between log of permeability and the difference in effective stress defined in Eq 3.110. The *points* are measurements

Based on Eqs. 3.100 and 3.105, log(k) is a linear function of the effective stress difference given in the above equation. Again, the agreement with data supports the theoretical results (Fig. 3.33).

Finally, it is important to indicate that the creeping processes are ignored in the data analysis, based on the consideration that permeability changes due to creeping are not expected to be significant in the experiments of Davy et al. (2007). For example, the laboratory experiments of Jobmann et al. (2010) showed that over about 5 days, fracture permeability was reduced by 20 % only for Opalinus clay. This permeability change is much smaller than those observed in the experiments of Davy et al. (2007) (Fig. 3.32), although Opalinus clay is softer than the Callovo-Oxfordian argillite rock studied in Davy et al. (2007) and therefore subject to a larger degree of creeping.

3.4 Coupled Hydro-mechanical Processes in a Dual-Continuum System

The coupled hydro-mechanical process in fractured rocks, as previously indicated, is important for many energy development and environment management applications. Mechanical deformation of subsurface materials generally affects fluid flow within them because their hydraulic properties are dependent on the deformation. On the other hand, effective stress distribution is influenced by the flow process that determines the corresponding pore-pressure distribution. This section, based on the work of Liu and Rutqvist (2013), discusses modeling of the coupled hydro-mechanical process associated with multiphase flow in fractured rock.

As indicated in Chap. 2, two major approaches are available in the literature for modeling flow and transport processes in fractured rocks: the discrete-fracture-network approach and the continuum approach. They differ in their manners to treat fracture networks in the model structure. The fracture-network approach explicitly considers individual fractures, their interaction with rock matrix, and flow process within them. Thus it often involves computational generation of synthetic fracture networks, based on observed fracture geometry data, and subsequent modeling of flow and transport in these networks. This approach is useful for detailed studies of small-scale flow processes and for dealing with situations where flow process is dominated by a small number of fractures or features (e.g., faults). However, it does have some limitations for practical applications. For example, it is very computationally intensive if all the individual fractures are considered for a large-scale problem. Furthermore, significant uncertainties exist in determining fracture geometry, fracture spatial distribution, and the associated flow properties, which may make the corresponding simulation results not very meaningful in some cases.

The continuum approach is essentially an upscaling approach and focuses on processes at a scale much larger than a typical length of fractures under consideration. Below that scale, fractures are considered to be sufficiently ubiquitous and distributed in such a way that they can be meaningfully described statistically (Barenblatt and Zheltov 1960; Warren and Root 1963). Thus, the continuum approach is a large-scale approach and based on a reasoning that large-scale processes may not depend on specific details of fracture geometry and distribution in space. In this case, connected fractures and rock matrix are viewed as two or more overlapping interacting continua and the continuum-mechanics formulations can be used to describe flow and transport in each continuum. Coupling of the processes between different continua is determined by their interaction mechanisms at a subgrid scale. Because the continuum approach is relatively simple and straightforward to implement, it is preferred for many applications encountered in practice (National Research Council 1996). Note that the above discussion about the fracture-network approach and the continuum approach is applicable to the mechanical deformation process as well.

The most commonly used continuum approach is the so-called dual continuum approach in which fracture and rock-matrix are treated as two overlapping continua. While the dual-continuum approach was initially developed for modeling flow process and has been widely used in the literature for both single-phase and multiphase flow systems (e.g., Barenblatt and Zheltov 1960; Warren and Root 1963; Kazemi and Gilman 1993; Liu et al. 1998), considerable effort has been made with respect to modeling mechanical deformation processes and their coupling with flow processes in fractured rock (e.g., Wilson and Aifantis 1982; Bai et al. 1993; Berryman and Wang 1995). These latter studies generally demonstrate the importance of dual-continuum behavior in flow and mechanical deformation processes resulting from dramatic differences between fracture and matrix (hydraulic and mechanical) properties. This section introduces the work of Liu and Rutqvist (2013) that is based on the TPHM and the dual-continuum approach. Considering that micro-fractures (and the alike) play an important role in the development of the TPHM, it may be useful to emphasize the difference between micro-fractures and the fractures mentioned here. It is relevant at least for some readers, given the fact that fractures exist at different scales. In general, micro-fractures are small fractures with a typical length on the order of micro meters and fractures correspond to those that we can see with our eyes from the surface of a fractured rock.

As noted by Liu and Rutqvist (2013), most of the previous studies of coupled hydro-mechanical processes in dual continuum systems are associated with single-phase flow process (e.g., Wilson and Aifantis 1982; Bai et al. 1993; Berryman and Wang 1995). Multiphase flow systems are generally more complex, because flow properties (such as permeability) depend on both stress and fluid saturations of different phases. To the best of our knowledge, Liu and Rutqvist (2013) reported one of the very few studies in the literature on modeling coupled hydro-mechanical processes in dual-continuum systems associated with multiphase flow process.

In commonly used coupled hydro-mechanical models (codes), as reviewed in Rutqvist et al. (2001), the coupling between hydraulic and mechanical deformation processes is implemented mainly through the mass conservations of the fluid component and solid phase. When fluid flow is of the most interest, it is more straightforward to study the coupling between fluid flow and rock deformation through a governing equation that focuses on the evolution of pore space (rather than solid phase) during the deformation, because fluid flow occurs in pore spaces (or within fractures). Liu and Rutqvist (2013) provided a derivation of such a governing equation that will be discussed later.

There are two numerical methods for modeling coupled hydro-mechanical processes, as discussed in a number of studies on single continuum systems (e.g., Settari and Mourits 1998; Settari and Walters 2001, Rutqvist et al. 2002; Mainguy and Longuemare 2002). One is called a "fully" coupled method that directly solves governing equations for the coupled processes within an individual code. The other is a "partially" coupled method that uses a conventional reservoir simulator in conjunction with a geomechanical simulator, and information exchanges between the two simulators. Both methods can accurately solve coupled hydro-mechanical processes. In addition to its ease of implementation, the latter method can take advantage of the fact that powerful and sophisticated simulators are available in both areas and their capabilities have kept expanding. [Studies on how to achieve numerically stable solutions to problems using the "partially" coupled method for flow and mechanical processes are also reported by Armero and Simo (1992), Armero (1999), Wheeler and Gai (2007) and Zienkiewicz et al. (1988).] Nevertheless, both methods require the development of formulations for coupled processes such that they allow for accurate information exchange between flow and mechanical deformation processes. This section will present the formulations (governing equations and constitutive relationships) for coupled hydro-mechanical processes associated with multiphase flow in deformable fractured rock (Liu and Rutqvist 2013).

3.4.1 Governing Equations

Governing equations, generally in a form of differential equations, are mathematical expressions of basic physical laws, such as mass balance and energy conservation. The governing equations for the coupled hydro-mechanical processes are based on mass conservation of fluid and solid phases (Rutqvist et al. 2001). This subsection derives an equation for the evolution of pore space during deformation, and then presents fluid mass balance equations for a dual-continuum system (Liu and Rutqvist 2013).

3.4.1.1 Pore-Space Evolution

Pore-space evolution refers to the pore-space change, as a function of time, within a control volume (V_c) with fixed boundaries in the deformed rock body. For simplicity, the rock is assumed to be homogeneous and isotropic. The pore-space evolution process can be described by that the increase in pore space (storage term) for the rock body within the control volume during a unit time is equal to the summation of net pore-space flux into the control volume due to solid flow (advection term) and local expansion of pore space (expansion term) within the control volume during that unit time.

It is easy to show that the storage and advection terms are $\frac{\partial(V_c \phi_j)}{\partial t}$ and $-V_c \nabla \cdot (\phi_j \boldsymbol{v}_{s,j})$, respectively, where ϕ_j is porosity for continuum j and defined as the ratio of pore volume for that continuum to the total (bulk) volume of rock body (including all the continua), t is time, and $\boldsymbol{v}_s$ is the solid velocity vector. A useful way to understand the above two terms is to consider the pore space to be a flowing fluid with a density of ϕ_j and a velocity of $\boldsymbol{v}_s$. If some readers cannot directly obtain the terms based on the similarity to fluid flow, they are encouraged to check the terms by deriving them for a relatively simple one-dimensional system.

The pore space evolution equation that describes the pore evolution process consists of storage, expansion and advection terms and will be presented later. While the storage and advection terms are given above, the rest of this subsection is to derive an expression for the expansion term in order to obtain the pore space evolution equation. During the following derivation, the small strain assumption, following Liu and Rutqvist (2013), will be used for simplicity. However, the derivation procedure can be extended to a more general case.

Liu and Rutqvist (2013) derived the expansion term from bulk volume change and solid-phase volume change for a unit rock volume. Using the definition of strain, bulk-volume change is given as

$$dV_j = V_{0,j} d\varepsilon_{v,j} \quad (j = f, m) \tag{3.111}$$

where V is the bulk rock volume, V_0 is the rock volume under the zero stress condition, and ε_v is the volumetric strain that is considered positive for expansion in this section. [Readers should keep in mind that in Sect. 3.4 the positive directions for both volumetric strain and stress are opposite to those in the other sections of this chapter such that the volumetric strain and porosity change have the same positive direction here (Liu and Rutqvist 2013).] The subscripts f and m refer to the fracture and matrix continuum, respectively. Similarly, solid volume change can be expressed as

$$dV_{s,j} = V_{s,j} d\varepsilon_{s,j} \tag{3.112}$$

where subscript s refers to the solid phase here. Using definitions of the solid and bulk rock volumes, respectively, one can obtain

$$V_{s,j} = \left(1 - \phi_j\right) V_{0,j} \tag{3.113}$$

Hooke's Law further gives

$$d\varepsilon_{s,j} = \frac{d\sigma_{s,j}}{K_{s,j}} \tag{3.114}$$

and

$$d\varepsilon_{v,j} = \frac{d\sigma_j}{K_j} \tag{3.115}$$

where $\sigma_{s,j}$ and $K_{s,j}$ are the (average) normal stresses acting on the solid phase and solid-phase modulus for the continuum j, respectively. The variables σ_j and K_j are respective (average) effective stress and bulk modulus for the continuum j. Considering that effective stress is defined with respect to the bulk rock body (including both solid phase and pore space), one can relate it to $\sigma_{s,j}$ by

$$d\sigma_j = \left(1 - \phi_j\right) d\sigma_{s,j} + \left(\phi_j - \alpha_j\right) dP_j \tag{3.116}$$

where P_j and $\alpha_j = 1 - \frac{K_j}{K_{s,j}}$ are the pore pressure and Biot's coefficient, respectively, in the continuum j. The above equation is derived from the relation $\sigma_j = \sigma_{t,j} - \alpha_j P_j = \left(1 - \phi_j\right)\sigma_{s,j} + \phi_j P_j - \alpha_j P_j = \left(1 - \phi_j\right)\sigma_{s,j} + \left(\phi_j - \alpha_j\right) P_j$, where $\sigma_{t,j}$ is the total stress for continuum j. Differentiating the above relation, while ignoring the porosity change (small strain assumption) yields Eq. 3.116.

Combining Eqs. 3.111–3.116 results in the following expansion term

$$\frac{\partial\left(V_j - V_{s,j}\right)}{\partial t} = \Phi_j V_c \left[\alpha_j \frac{\partial \varepsilon_{v,j}}{\partial t} - \frac{\left(1 - \alpha_j\right)\left(\phi_j - \alpha_j\right)}{K_j} \frac{\partial P_j}{\partial t}\right] \quad (j = f, m) \tag{3.117}$$

where $\Phi_j = \frac{V_{0,j}}{V_c}$ is the volume fraction of jth continuum (in the control volume) under the unstressed condition, and the volume difference $V_j - V_{s,j}$ is equal to the pore volume for that continuum within the control volume.

Based on the obtained mathematical expressions for all three terms (storage, advection and expansion), the pore-space evolution equation can be given as (Liu and Rutqvist 2013)

$$\frac{\partial \phi_j}{\partial t} + \nabla \cdot (\phi_j \boldsymbol{v}_{s,j}) = \Phi_j \left[\alpha_j \frac{\partial \varepsilon_{v,j}}{\partial t} - \frac{(1-\alpha_j)(\phi_j - \alpha_j)}{K_j} \frac{\partial P_j}{\partial t} \right] \quad (j = f, m) \quad (3.118)$$

It establishes the fundamental relationship among porosity, solid-phase flow and local volumetric strain during rock deformation processes. While the derivation has focused on a homogeneous and isotropic rock body, it can be extended to more complex situations. Note that a solid phase for the fracture continuum does not exist because a fracture continuum consists of fracture voids only. Thus a consideration of solid-phase volume change is not required for deriving a pore-space expansion term for the fracture continuum. Accordingly, Biot's coefficient (α_f) should be equal to one (Berryman and Wang 1995). In other words, the use of $\alpha_f = 1$ in the above equation gives the correct result for the fracture continuum.

3.4.1.2 Fluid Mass Conservation

Fluid mass conservation equations are the governing equations for modeling flow processes. In its general form, a fluid mass conservation equation for multiphase flow is given as (e.g., Olivella et al. 1994; Pruess 1991; Rutqvist et al. 2001; Liu and Rutqvist 2013)

$$\frac{\partial M_{j,\psi}^k}{\partial t} + \nabla \cdot \boldsymbol{q}_{j,\psi}^k - Q_{j,\psi}^k = 0 \quad (3.119)$$

where $M_{j,\psi}^k$ is the fluid mass with jth continuum per unit control volume of fractured rock, superscript k is the fluid component (such as water, air, or CO_2), subscript ψ is the phases (gas, liquid, or solid). The symbol $\boldsymbol{q}_{j,\psi}^k$ represents fluid flux (within the jth continuum) with respect to the spatial reference system, and $Q_{j,\psi}^k$ is the production rate of component k per unit volume. The above equation simply states that for a given fluid component in a given phase and continuum, the rate of increase in its mass within a control volume is equal to a combination of the net fluid flux into the control volume and the production rate within it.

In Eq. 3.119, the mass $M_{j,\psi}^k$ and mass flux $\boldsymbol{q}_{j,\psi}^k$ can be expressed as

$$M_{j,\psi}^k = \phi_j S_{j,\psi} \rho_{j,\psi}^k \quad (3.120)$$

and

$$\boldsymbol{q}_{j,\psi}^{k} = \phi_j S_{j,\psi} \rho_{j,\psi}^{k} \left(\boldsymbol{v}_{j,r\psi} + \boldsymbol{v}_{s,j} \right) = \boldsymbol{q}_{j,r\psi}^{k} + \phi_j S_{j,\psi} \rho_{j,\psi}^{k} \boldsymbol{v}_{s,j} \tag{3.121}$$

where $S_{j,\psi}$ is the phase saturation for the jth continuum, $\rho_{j,\psi}^{k}$ is mass of the fluid component k per unit volume of the phase, $\boldsymbol{v}_s$ again is the solid velocity, and $\boldsymbol{q}_{j,rw}^{k}$ is fluid flux relative to the solid. The relative velocity is defined by

$$\boldsymbol{v}_{j,r\psi} = \boldsymbol{v}_{j,\psi} - \boldsymbol{v}_{s,j} \tag{3.122}$$

where $\boldsymbol{v}_{j,\psi}$ is the velocity of the fluid phase with respect to a fixed reference system.

As demonstrated in Chap. 1, Darcy's law describes fluid flow when solid phase is fixed. When the solid phase involves deformation, it should be applied to describe $\boldsymbol{q}_{j,rw}^{k}$, rather than $\boldsymbol{q}_{j\psi}^{k}$. As a result, we need to derive the fluid mass balance equation in terms of $\boldsymbol{q}_{j,rw}^{k}$ that is generally used in reservoir simulators for modeling multiphase flow within rigid media. Combining Eqs. 3.119–3.122 yields

$$\frac{\partial \left(M_{j,\psi}^{k} \right)}{\partial t} + \nabla \cdot \boldsymbol{q}_{j,r\psi}^{k} + \nabla \cdot \left(\phi_j S_{j,\psi} \rho_{j,\psi}^{k} \boldsymbol{v}_{s,j} \right) - \boldsymbol{Q}_{j,\psi}^{k} = 0 \tag{3.123}$$

Expanding the third term in the above equation and using the pore-space evolution equation (Eq. 3.118) gives

$$\begin{aligned} \nabla \cdot \left(\phi_j S_{j,\psi} \rho_{j,\psi}^{k} \boldsymbol{v}_{s,j} \right) &= S_{j,\psi} \rho_{j,\psi}^{k} \nabla \cdot \left(\phi_j \boldsymbol{v}_{s,j} \right) + \phi_j \boldsymbol{v}_{s,j} \cdot \nabla \left(S_{j,\psi} \rho_{j,\psi}^{k} \right) \\ &\approx \frac{M_{j,\psi}^{k}}{\phi_j} \left[\Phi_j \alpha_j \frac{\partial \varepsilon_{v,j}}{\partial t} - \Phi_j \frac{\left(1 - \alpha_j\right)\left(\phi_j - \alpha_j\right)}{K_j} \frac{\partial P_j}{\partial t} - \frac{\partial \phi_j}{\partial t} \right] \end{aligned} \tag{3.124}$$

Note that we ignore the small term $\phi_j \boldsymbol{v}_{s,j} \cdot \nabla \left(S_{j,\psi} \rho_{j,\psi}^{k} \right)$ (owing to small strain or small solid velocity) in the above derivation. Inserting Eq. 3.124 into Eq. 3.123 results in the final form of the fluid mass-balance equation of Liu and Rutqvist (2013):

$$\begin{aligned} &\frac{\partial \left(M_{j,\psi}^{k} \right)}{\partial t} + \nabla \cdot \boldsymbol{q}_{j\psi}^{k} + \frac{M_{j,\psi}^{k}}{\phi_j} \left(\Phi_j \alpha_j \frac{\partial \varepsilon_{v,j}}{\partial t} - \Phi_j \frac{\left(1 - \alpha_j\right)\left(\phi_j - \alpha_j\right)}{K_j} \frac{\partial P_j}{\partial t} - \frac{\partial \phi_j}{\partial t} \right) - \boldsymbol{Q}_{j,\psi}^{k} \\ &= 0 \end{aligned} \tag{3.125}$$

The third term in Eq. 3.125 is an additional term owing to mechanical deformation, as compared with the fluid mass balance equations commonly used in reservoir simulators without considering the effect of mechanical deformation (e.g., Pruess 1991). The relative importance of this term, which will be evaluated in a sensitivity study to be discussed later, determines whether the term needs to be included for modeling coupled hydro-mechanical processes.

Equation 3.125 can be further simplified in some special cases. For example, when solid phase is incompressible, the corresponding Biot's coefficient (α_j) is equal to one. Biot's coefficient characterizes the degree of coupling between hydraulic and mechanical processes. A larger Biot's coefficient corresponds to a larger degree of coupling. For a zero value of the Biot's coefficient, hydraulic and mechanical processes are not coupled, but rather remain independent. A simple relationship between volumetric strain and porosity can be obtained for $\alpha_j = 1$.

For a fractured rock with an incompressible solid phase, one has

$$d\varepsilon_{v,j} = \frac{dV_j}{V_{0,j}} = \frac{dV_j^p}{V_0 \Phi_j} \tag{3.126}$$

where superscript p again refers to pore space. By definition, porosity can be related to pore volume $V_{p,j}$ by

$$d\phi_j = d\left(\frac{V_j^p}{V}\right) = \frac{1}{V}\left(dV_j^p - \phi dV\right) \tag{3.127}$$

Combining Eqs. 3.126 and 3.127 and using $\frac{V}{V_0} \approx 1$ (owing to small strain) yields

$$\Phi_j d\varepsilon_{v,j} - \phi_j d\varepsilon_v = d\phi_j \tag{3.128}$$

where $d\varepsilon_v = \frac{dV}{V_0}$ is total volumetric strain for bulk fractured rock (including both fracture and matrix continua). Applying the procedure to derive Eq. 3.128 to the bulk rock body, one can obtain:

$$(1 - \phi)d\varepsilon_v = d\phi \tag{3.129}$$

where ϕ is the total porosity for fractured rock. Based on the above two equations, Eq. 3.125 can be simplified to (Liu and Rutqvist 2013)

$$\frac{\partial\left(M_{j,\psi}^k\right)}{\partial t} + \nabla \cdot \boldsymbol{q}_{j,r\psi}^k + \frac{M_{j,\psi}^k}{1 - \phi}\left(\frac{\partial \phi}{\partial t}\right) - \boldsymbol{Q}_{j,\psi}^k = 0 \tag{3.130}$$

Note that the third term (the additional storage term) on the left hand side of the above equation is actually dependent on the total porosity, rather than the porosity of the corresponding continuum.

3.4.1.3 Momentum Conservation

The momentum-conservation equation is a fundamental relation used for simulating rock deformation processes. While details regarding derivation of this equation are available in the literature (e.g., Jaeger et al. 2007; Fjær et al. 2008), it is presented here just for the sake of completeness. In the absence of an inertia term (commonly assumed for modeling coupled hydro-mechanical processes), the momentum conservation can be expressed as (Rutqvist et al. 2001)

$$\nabla \cdot \boldsymbol{\sigma}_T + \rho_b \mathbf{g} = 0 \tag{3.131}$$

where $\boldsymbol{\sigma}_T$ is the total stress tensor, $\mathbf{g}$ is the acceleration vector for gravity, and ρ_b is the bulk density including contributions from both fluid and solid phases.

3.4.2 *Constitutive Relationships*

In addition to the governing equations, constitutive relationships (between hydraulic parameters, mechanical parameters, and related variables) are needed for modeling the coupled hydro-mechanical process. While constitutive relationships for uncoupled hydraulic and mechanical processes have been very well documented in the literature (e.g., Pruess 1991; Jaeger et al. 2007), this subsection presents the TPHM-based relationships for the coupled hydro-mechanical process in a dual-continuum system (Liu and Rutqvist 2013).

3.4.2.1 Stress-Dependent Hydraulic Properties

The coupling between hydraulic and mechanical processes is largely determined by the stress dependence of hydraulic properties, which is closely related to the stress-strain relationships in fractured rock. Based on Eq. 3.28, the bulk modulus K_j for the jth continuum is given by

$$K_j = \frac{1}{\frac{\gamma_{e,j}}{K_{e,j}} + \frac{\gamma_{t,j}}{K_{t,j}} \exp\left(-\frac{\sigma_j}{K_{t,j}}\right)} \tag{3.132}$$

where subscripts 0, e, and t again denote the unstressed state, the hard part and the soft part, respectively, for a rock body, and K_e and K_t refer to bulk moduli for the two parts. The constants $\gamma_{e,j}$ and $\gamma_{t,j}$ represent volumetric fractions of the hard and soft parts under the unstressed condition $\left(\gamma_{e,j} + \gamma_{t,j} = 1\right)$.

The stress-dependent matrix porosity is given by Eq. 3.26 or Eq. 3.27. Once the matrix porosity is known for a given stress, matrix permeability can be estimated through the relations between permeability and porosity (Sect. 3.2.2.4).

The relationship for the stress-dependence of fracture aperture (for a single fracture) (Eq. 3.79) allows for relating fracture aperture b (and fracture porosity) to the effective normal stress by

$$\frac{b}{b_0} = \frac{\phi_f}{\phi_{0,f}} = \gamma_{e,f}\left(1 - \frac{\sigma_f}{K_{e,f}}\right) + \gamma_{t,f}\exp\left(-\frac{\sigma_f}{K_{t,f}}\right) \qquad (3.133)$$

In the above equation, the stress-dependent behavior of fracture aperture is controlled by the second term at a low stress and the first term at a high stress on the right hand side. There is a considerable amount of laboratory data indicating that fracture closure (or fracture aperture) remains practically unchanged at high stress (e.g., Goodman 1976; Barton et al. 1985). This is equivalent to saying that in practice the following condition holds:

$$1 - \frac{\sigma_f}{K_{e,f}} \approx 1 \qquad (3.134)$$

In this case, Eq. 3.133 can be reduced to (Liu and Rutqvist 2013):

$$\frac{b}{b_0} = \frac{\phi_f}{\phi_{0,f}} = \gamma_{e,f} + \gamma_{t,f}\exp\left(-\frac{\sigma_f}{K_{t,f}}\right) \qquad (3.135)$$

Then stress-dependence of fracture permeability can be estimated by the cubic law that states that fracture permeability is proportional to the cube of the fracture aperture (a function of stress), as discussed in Sect. 3.3.1.1. However, Eqs. 3.133–3.135 are developed by considering normal stress only and should be used when fracture planes are perpendicular to principal stresses.

Note that Liu and Rutqvist (2013) do not yet consider the stress-dependent multiphase flow properties, such as capillary pressure and relative permeability. The stress dependence of these multiphase flow properties is already discussed in detail in Sect. 3.3 and may be combined with the work of Liu and Rutqvist (2013) for future studies.

3.4.2.2 Effective Parameters for Dual Continua

A geomechanical simulator generally models bulk deformation (including both fractures and rock matrix). When the "partially" coupled method is used, it is necessary to develop expressions of effective parameters for such a dual-continuum system such that appropriate information transfer can occur between geomechanical and reservoir simulators.

One important parameter for modeling the coupled hydro-mechanical process is the effective stress that depends on both total normal stress, σ_T, and fluid pressure. For a dual-continuum system under hydrostatic stress conditions (with the total stress assumed to be the same in both continua as a requirement for continuity of stress), one has

$$d\varepsilon_{v,m} = \frac{dV_m}{V_{0,m}} = \frac{d(\sigma_T + \alpha_m P_m)}{K_m} \tag{3.136}$$

$$d\varepsilon_{v,f} = \frac{dV_f}{V_{0,f}} = \frac{d(\sigma_T + \alpha_f P_f)}{K_f} \tag{3.137}$$

$$d\varepsilon_v = \frac{dV}{V_0} = \frac{dV_m + dV_f}{V_0} = \Phi_f d\varepsilon_{v,f} + \Phi_m d\varepsilon_{v,m} = \frac{d\sigma_{eff}}{K_{eff}} \tag{3.138}$$

Combining Eqs. 2.136–2.138, Liu and Rutqvist (2013) obtained the effective modulus K_{eff} and stress σ_{eff} for a dual-continuum system given by

$$K_{eff} = \frac{1}{\frac{\Phi_f}{K_f} + \frac{\Phi_m}{K_m}} \tag{3.139}$$

$$\sigma_{eff} = \sigma_T + \frac{\Phi_f K_{eff}}{K_f} P_f + \frac{\Phi_m \alpha_m K_{eff}}{K_m} P_m \tag{3.140}$$

Effective parameters (Eqs. 3.139 and 3.140) are directly used in a geomechnical simulator to model deformation processes. Equation 3.140 is reduced to Eq. 3.99 for a single continuum system with $\Phi_f = 0$, $\Phi_m = 1$, and $K_{eff} = K_m$. (Note that Eqs. 3.99 and 3.140 have the opposite positive directions for the stress.) Thus, Eq. 3.140 can be considered to be a generalization of its counterpart for the single continuum (Eq. 3.99).

As previously indicated, Eqs. 3.139 and 3.140 are developed under the uniform stress condition. The same condition has also been used in previous studies of dual-continuum mechanical behavior (e.g., Berryman and Wang 1995). Consequently, the results here are essentially identical to those reported in Berryman and Wang (1995). When the total stress is direction dependent, effective stress σ_{eff} in a principal-stress direction is calculated using the total stress in that direction. On the other hand, Eqs. 3.139 and 3.140 are developed by considering normal stress only and therefore valid when fracture planes are perpendicular to principal stresses. More studies are needed for developing rigorous approaches to dealing with more complex fracture networks.

The simulation procedure for the "partially" coupled method, using formulations presented in Sect. 3.4, can be briefly described as follows. For an assumed pressure distribution, effective parameters are calculated with Eqs. 3.139 and 3.140. Then, the corresponding geomechanical simulator is run to obtain distributions for stresses and strains. Effective stresses and strains for the two continua are evaluated using Eqs. 3.136, 3.137 and 3.140. The stress-dependent hydraulic properties are updated using relationships described in Sect. 3.4.2.1. The fluid mass balance equations are solved using a reservoir simulator with updated parameters to obtain flow fields and fluid pressure distributions. With the new pressure distributions, effective parameters are updated and the corresponding geomechanical simulator is rerun. This iteration procedure continues until convergence is achieved for a given time step.

3.4.3 An Application to the In Salah CO_2 Injection Project

CO_2 geological sequestration is a technique for the long-term storage of CO_2 in deep geological formations to mitigate or defer global warming; the CO_2 (stored by the technique) is either removed from the atmosphere or other sources that otherwise would release the CO_2 into the atmosphere. The technique involves injection of supercritical CO_2 into a deep formation that disturbs the ambient pore pressure distribution there and induces effective stress changes. As a result, the coupled hydro-mechanical process plays an important role in almost all the CO_2 geological sequestration projects.

Sections 3.4.1 and 3.4.2 present a systematic approach to modeling coupled hydro-mechanical process in a dual-continuum system. In this subsection, we demonstrate usefulness of the approach by showing the consistence between modeling results with data from a CO_2 injection test that is part of the In Salah Project. The In Salah Project, located in the central region of Algeria, is the world's first on-shore industrial scale CO_2 storage project. Natural gas produced from the area is high in CO_2 and that CO_2 is being returned to the earth for geological storage. Nearly one million tonnes of CO_2 per year has been injected since August 2004 into relatively low-permeability, 20 m thick, water-filled carboniferous sandstone at a depth of about 1800–1900 m, around the Krechba gas field (Fig. 3.34). To ensure adequate CO_2 flow-rates across the low-permeability sand-face, the In Salah Gas Project decided to use long-reach (about 1–1.5 km) horizontal injection wells (Wright 2008). The storage formation is an excellent

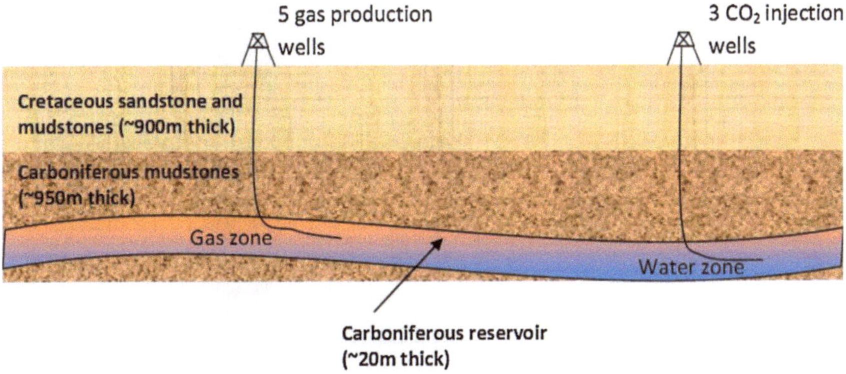

Fig. 3.34 A schematic cross section of In Salah site

analogue for much of North-West Europe and the U.S. Midwestern region, where large CO_2-storage will be required if CO_2 capture and geological storage is to make a significant contribution to addressing CO_2 emissions. The formation where CO_2 injection has occurred is known to contain well-connected fractures.

Based on the dual-continuum approach presented in Sects. 3.4.1 and 3.4.2, Liu and Rutqvist (2013) developed a numerical model of CO_2 injection for the In Salah Gas project. To the best of our knowledge, the work probably represents the first effort to use a dual-continuum approach for modeling coupled processes in a CO_2 sequestration problem. It is also important to emphasize that the focus of the model is on evaluating the relative importance of the additional storage term in Eq. 3.130, rather than on the detailed hydro-mechanical processes involved in the CO_2 geological sequestration of the In Salah Gas project.

Because the formation for the CO_2 injection is relatively thin (20 m thick), a two-dimensional fine-grid model was developed to represent this formation (Fig. 3.35). Each gridblock contains a fracture and a matrix element. Fluid flow occurs through fractures and between fractures and the rock matrix. Relatively small grid spacing (20 m) is employed near the injection well, and constant-pressure boundary conditions were used on the four sides of the model domain. The model domain includes the gas reservoir treated as a constant-pressure body. This is justified because pressure in the gas reservoir is not expected to change dramatically during the period of CO_2 injection as a result of the large volume of the gas reservoir. Hydraulic and mechanical properties are determined from laboratory and field measurements. The matrix porosity ($\phi_{e,m}$) and permeability are 17 % and 5 md, respectively, and the matrix bulk modulus is 9 GPa. Fracture properties are calibrated using well-injection data (to be discussed later), while the matrix properties remain fixed during calibration, given that reliable laboratory measurements are available for these parameters.

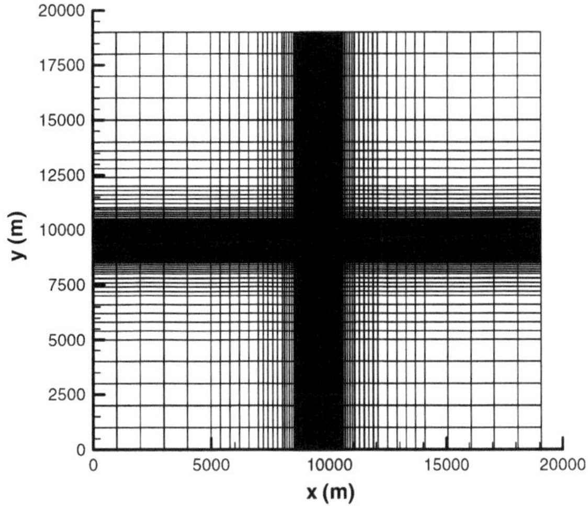

Fig. 3.35 The two-dimensional numerical grid. The injection well is located in the center of the grid (Liu and Rutqvist, 2013)

For this simplified two-dimensional model for the injection formation, Liu and Rutqvist (2013) further assumed the total stress to be unchanged during CO_2 injection processes. In this case, changes in effective stress are totally determined by changes in fluid pressure. A recent study seems to indicate that this assumption may be reasonable for determining a first-order estimation of maximum sustainable injection pressure (associated with CO_2 geological sequestration), although a more accurate consideration of coupling between fluid flow and geomechanical processes is desirable (Liu and Rutqvist 2013). Note that this assumption will eliminate the need to directly couple the reservoir simulator with a geomechanical simulator, as long as the reservoir simulator incorporates the new governing equation for fluid flow (e.g., Eq. 3.124) and stress-dependent hydraulic properties. The reservoir simulator TOUGH2 (Pruess 1991) was used in this study. TOUGH2 is a general numerical simulator for flow and transport in porous and fractured media and includes a module (ECO_2N) specifically developed for modeling CO_2 geological sequestration in deep aquifers. In the study of Liu and Rutqvist (2013), TOUGH2 is modified to consider the additional storage term in the governing equation for fluid flow and stress dependence of hydraulic properties.

The numerical model is calibrated against wellhead pressure (WHP) data observed from the In Salah Project site (Fig. 3.36). The pressure difference in Fig. 3.36 is defined as the difference between WHP and its value corresponding to the zero injection. In this part of the study, Liu and Rutqvist (2013) considered this difference to be the same as bottom hole pressure (BHP) difference. The measured flow rates at the injection well are given in Fig. 3.37. During model calibration, the measured flow rates are approximated by a step function (Fig. 3.37) because the data is subject to a large-degree of temporal fluctuation. The calibrated parameters are fracture permeability under the ambient condition and the mechanical properties given in Eq. 3.135. In general, injection pressure data before 200 days are used for

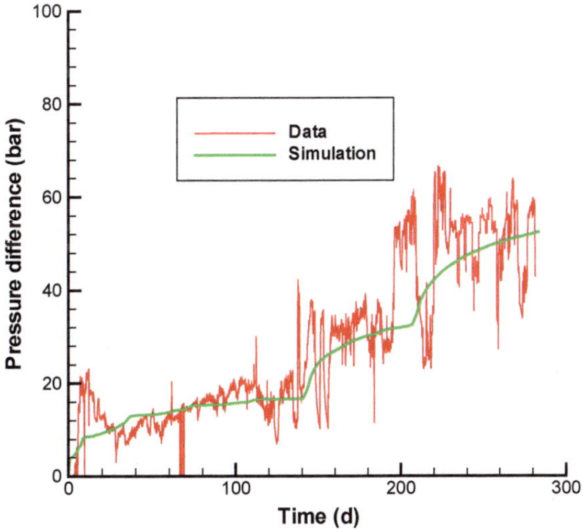

Fig. 3.36 Comparison between simulated and observed pressure differences (Liu and Rutqvist 2013)

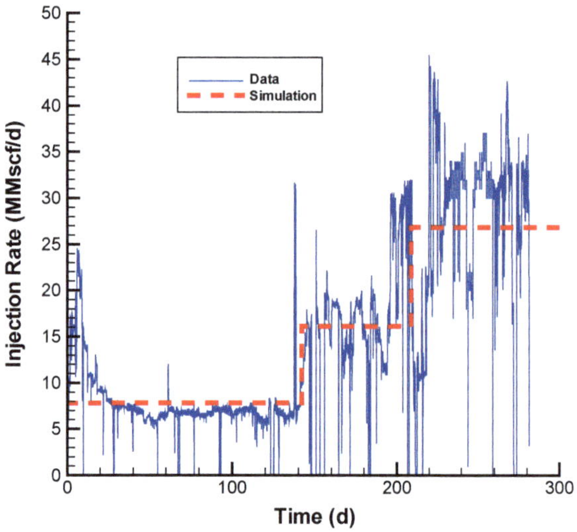

Fig. 3.37 The measured injection rates at the injection well. The *dashed line* is used in modeling study as an approximation of injection rate data (Liu and Rutqvist 2013)

determining ambient fracture permeability, because stress-dependent behavior becomes more pronounced for high-pressure differences at later time. Calibrated fracture permeabilities are 35 and 7 md, respectively, in the y and x directions (Fig. 3.35). During the calibration, the permeability ratio of 5 is assigned, and fracture porosity under ambient conditions is assumed to be 1 %. Calibrated fracture mechanical properties (Eq. 3.135) are $\frac{\gamma_{t,f}}{\gamma_{e,f}} = \frac{1}{8}$ and $K_{t,f} = 3$ MPa and consistent with the fracture property values given in Sect. 3.2. The calibration results capture the observed pressure changes (Fig. 3.36).

The calibrated model is then used for predicting injection pressure (at the injection well) as a function of time. The injection rate is assumed to be constant for the next 20 years and the same as the injection rate averaged over the last 100 days of the time period used for model calibration (Fig. 3.45). After 20 years, injection stops. The simulated injection pressures are presented in Fig. 3.38. Pressure initially increases and then decreases with the spreading of CO_2 plume. This pressure decrease is largely caused by the impact of the gas reservoir located on the left side of the injection well under consideration, and about 500 m away (in the x direction) from the well. When the CO_2 plume approaches the gas reservoir, the pressure increase can be significantly adsorbed by the gas reservoir that is assumed to have a constant pressure during the CO_2 geological sequestration period.

Note that simulated results with and without considering the storage term (in Eq. 3.130) are almost identical, suggesting that this new term has little effect on simulated fluid flow processes for this particular problem. Liu and Rutqvist (2013) examined this issue for a number of simulation scenarios and obtained similar comparison results. One particular scenario is that the matrix porosity decreases by 70 % and injection rate increases by 100 %, as compared with those used for generating simulation results in Fig. 3.38. These parameter changes are expected to

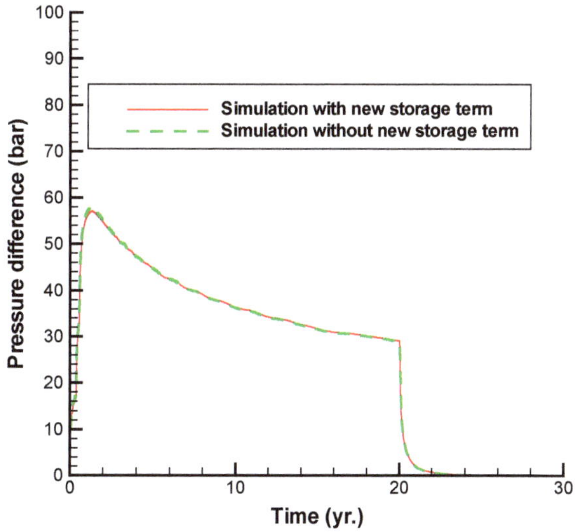

Fig. 3.38 Simulated injection pressures with and without considering the new storage term in the governing equation for fluid flow (Eq. 3.130) (Liu and Rutqvist 2013)

enhance the coupling between fluid flow and mechanical deformation for the fracture continuum, but still result in negligible differences between simulation results. However, more studies for different systems are needed to further confirm this finding, while the additional storage term can be easily incorporated into the currently existing reservoir simulators.

3.5 A Case Study: The Use of the TPHM to Model a Mine-by Test at Mont Terri Site, Switzerland

Shale has been considered as a potential host rock for geological disposal of high-level radioactive waste in several countries, as mentioned in Sect. 1.7.1. The key technical issue associated with a shale repository is the formation and evolution of the excavation damaged zone (EDZ) around repository, because the EDZ has higher permeability than intact rock, as a result of excavation-induced fractures, and therefore may act as the fast path for the leakage of the nuclear waste contaminations (Tsang et al. 2012). The mine-by (MB) test is a field-scale test conducted at the Mont Terri underground rock laboratory near the town of St. Ursanne in the Jura Mountains of northwestern Switzerland. The test objectives include investigating the EDZ behavior and the associated coupled hydro-mechanical processes in the Opalinus Clay during the excavation (Bossart and Thury 2007; Vietor 2011). MB test data sets provide a unique opportunity for validating numerical models. As a case study, this section briefly introduces the MB test and presents comparisons between simulation results and field measured data from the MB test (Bossart and

Thury 2007) to demonstrate the TPHM's capability to predict the dynamic hydro-mechanical properties of the rock mass around the excavation. This section is mainly based on the work of Li et al. (2014).

3.5.1 Mine-by (MB) Test at the Mont Terri Site and Numerical Model

The layout of the Mont Terri rock laboratory is shown in Fig. 3.39. Gallery 98 is the first dedicated research gallery there and was excavated in 1998. It was then extended in 2004 and the extension is called Gallery 04. The Mont Terri laboratory was expanded, at the end of 2006, by adding another gallery (called Gallery 08) of 165 m in length and 4 niches (called MB niches numbered 1 through 4) whose plan view is shown in Fig. 3.40a. An experiment team from several international organizations interested in shale repository, including Nagra, Andra, GRS, BRG, and Obayashi, took the opportunity to set up the MB test to investigate the deformation and the coupled hydro-mechanical behavior of the Opalinus Clay around bedding-parallel excavations (Bossart and Thury 2007; Vietor 2011). Sensors of different types were installed around MB Niche 2 (as shown in Fig. 3.40b), before it was drilled, to capture the hydraulic and mechanical effects of the excavation (Vietor 2011). The test focus was on acquiring data on the deformation and pressure evolution in the near field of MB Niche 2. The collected data sets are then used for improving the understanding of the processes leading to permanent damage of the clay rock and for model development and validation.

The main concern of this case study is the near-field responses around MB niche 2, including the deformation and hydraulic response to the excavation. As a 24 m long dead-end side niche of Gallery 08 with a diameter of 4.5 m, MB niche 2 was excavated from October 13th to November 7th in 2008 by a road header. Anchors and light shotcrete were chosen as permanent support (as shown in Fig. 3.41). To make sure that the test results are applicable, as much as possible, to a high-level waste emplacement tunnel, the niche was constructed in a similar manner for constructing the waste tunnel using the circular geometry of the niche section, small deviations from intended face geometry (vertical, without rounded corners), a constant excavation rate of 1.3 m/day with small variations, full-face excavation, and the same activity pattern every day. The resulting data are available for comparisons with modeling results for the niche deformation and the associated hydro-mechanical coupled processes during the excavation.

The excavation chronology of MB niche 2 is shown in Fig. 3.42. The total excavation period lasted over four weeks. For each sequential excavation and support step, support measures were usually installed in the morning, followed by renewed excavation in the afternoon. The pore pressure and deformation around the niche were monitored at selected locations with sensors (Fig. 3.40b). This excavation chronology is incorporated in the model study discussed in this section.

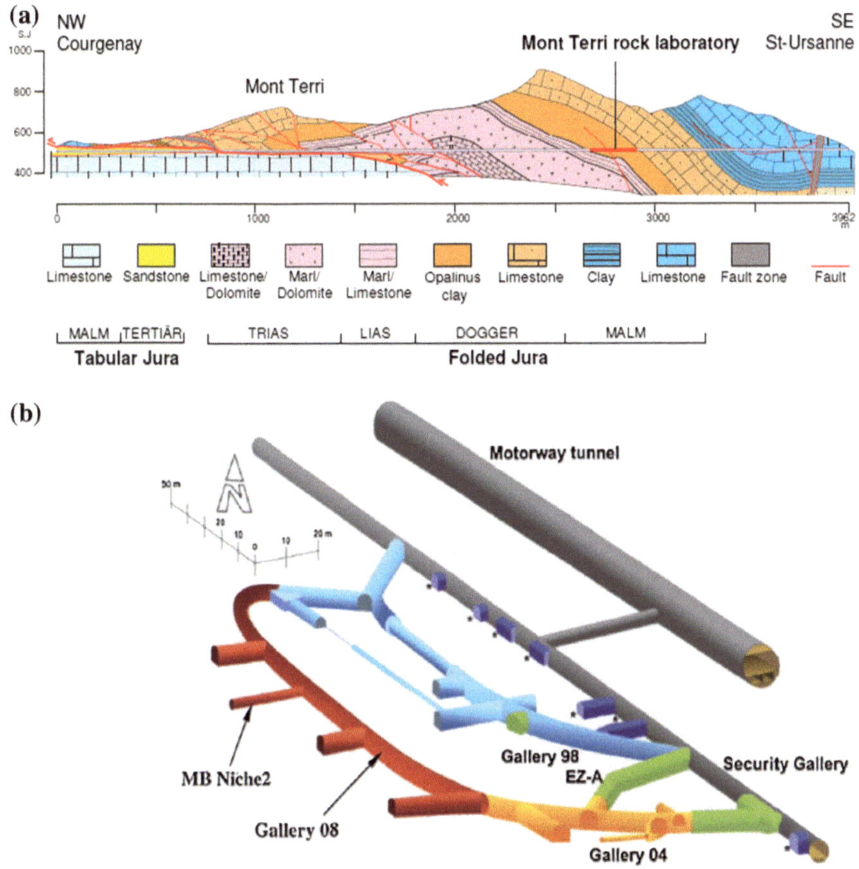

Fig. 3.39 The layout of the Mont Terri rock laboratory, **a** vertical cross-section of the Mont Terri anticline along the motorway tunnel, and **b** the general view of the layout of the Mont Terri rock laboratory (Bossart and Thury 2007) (reproduced by permission of Swiss Geological Survey at Swisstopo)

The geomechanical simulator, FLAC3D code (Itasca Consulting Group 2005) was used for all the numerical simulations for the case study. The TPHM was implemented into FLAC3D as the constitutive model here (Sect. 3.2.5). As previously indicated, the simulation results are compared with data sets from the MB tests, including a data set of changes in pore pressure as a function of time at selected locations around the niche. Change in pore pressure, P, is related to change in volumetric fluid content and volumetric strain, ε_v, by (Jaeger et al. 2007):

$$\frac{\partial \varsigma}{\partial t} = \frac{1}{M_B}\frac{\partial P}{\partial t} + \alpha_B \frac{\partial \varepsilon_v}{\partial t} \tag{3.141}$$

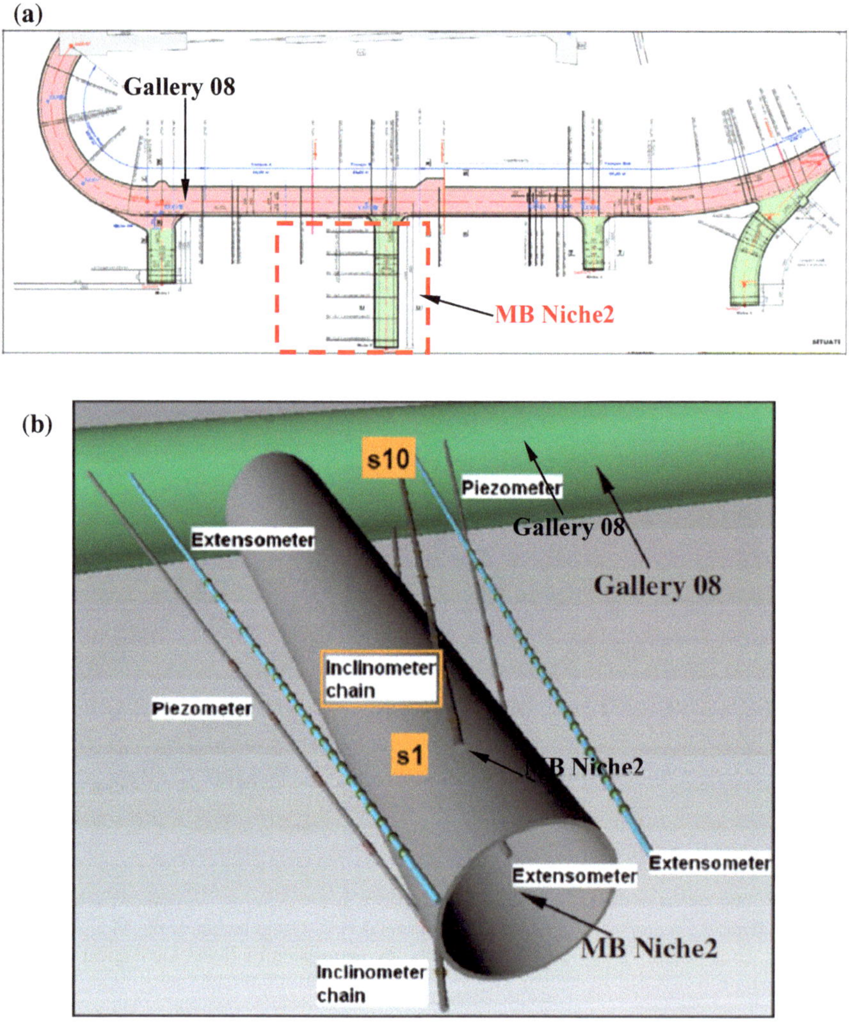

Fig. 3.40 The layout of MB Niche 2, **a** Gallery 08 and MB Niche 2 as planned (status Nov 2008) and **b** the instrumentation near MB Niche2 (Vietor 2011) (reproduced by permission of Nagra)

where ζ is the variation in fluid content or variation in fluid volume per unit volume of porous material due to diffusive fluid mass transport, as introduced by Biot (1956), and M_B is Biot's modulus. The above equation is derived based on the water mass balance.

Fig. 3.41 MB Niche 2 in construction: **a** the excavation advancing face, and **b** the view of the niche after reinforcement with shotcrete (Vietor 2011) (reproduced by permission of Nagra)

Fig. 3.42 The excavation chronology of MB niche2 (Li et al. 2014) (reproduced by permission of Elsevier)

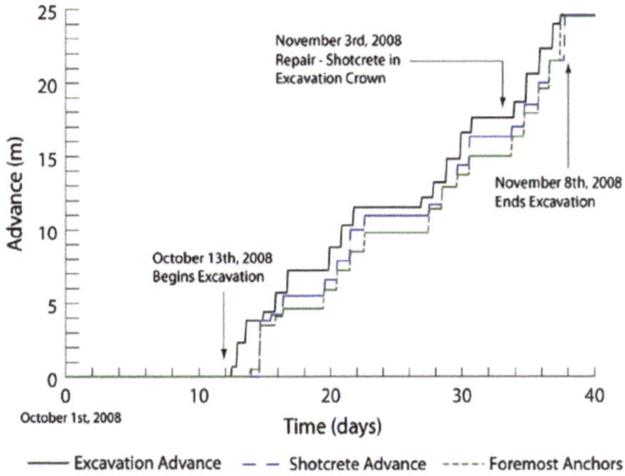

For a porous material, the Biot's modulus, M_B, is related to the fluid bulk modulus K_{fl} (Jaeger et al. 2007):

$$M_B = \frac{K_{fl}}{\phi + (\alpha_B - \phi)(1 - \alpha_B)K_{fl}/K_{ss}} \tag{3.142}$$

where again ϕ is the porosity and K_{ss} is the bulk modulus of solid skeleton. Within the TPHM framework, the relationship between the porosity and principal stresses (Eq. 3.59) contains the compressibility for the hard fraction of pore volume, C_e, that is selected as 2.0×10^{-9} Pa^{-1} here (Nagra 2002). The impact of porosity change on permeability is small for this case study and thus not considered for modeling the MB test. Also, the long-term response (creep or delayed plastic response) was not considered either in the model development, given the fact that the MB experiment only lasted a relatively short time period (Li et al. 2014).

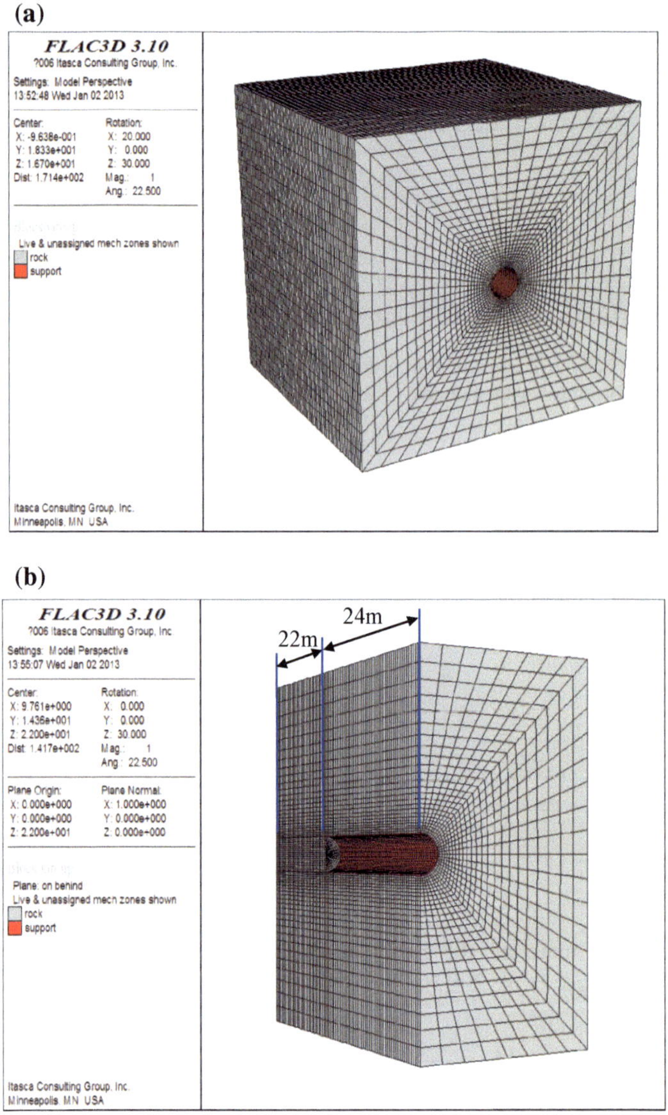

Fig. 3.43 The mesh of the numerical model. **a** The model mesh. **b** A section plane along Y-axis

The FLAC3D grid for MB niche 2, with a total of 100,256 zones (grid cells), is shown in Fig. 3.43 and has dimensions 40 m × 46 m × 44 m. Located in the center of the model is the MB niche 2 that is represented by a 24 m long tunnel. The longitudinal axis of the niche is located 22.0 m below the upper surface of the model. A system of coordinate axes is defined with the origin at the center of the tunnel, the z-axis in the upward direction and the y-axis along the axis of the niche.

Table 3.9 Physico-mechanical parameters used in simulations (Li et al. 2014)

Parameter	Value
Rock bulk density	2450 kg/m^3
Young's modulus for the hard part	4.0 GPa
Poisson's ratio	0.27
Tensional strength	1.2 MPa
Cohesion	4.0 MPa
Friction angle	24.0°
Young's modulus for the soft parts	2.0 MPa
γ_t	0.02
Initial porosity	0.137
Permeability for the intact rock mass	2e−20 m^2
Biot's coefficient	0.6
Fluid bulk modulus	1000 MPa
Fluid density	1000 kg/m^3
Fluid viscosity	1e−3 kg/(m s)

The inputs into the numerical model include hydraulic and mechanical properties for the rock formation and related components of the tunnel and also include operation parameters related to the excavation process, such as the excavation chronology. MB tests were conducted in the Opalinus Clay formation. Because Opalinus Clay is of interest as both a potential host formation for the radioactive waste repository and a formation through which new transportation tunnels have been constructed, a number of experimental investigations into values for hydraulic and mechanical parameters (obtained from laboratory experiments and field research) have been published by different groups (Corkum and Martin 2007b; Zhang et al. 2008; Bock 2009; Bock et al. 2010; Jobmann et al. 2010), as shown in Table 3.9. These parameter values are employed in this study.

The rock permeability value given in Table 3.9 corresponds to the intact rock mass. The shale matrix is assumed to be homogenous and fully saturated. The observed pore pressure is 2.0 MPa in the undisturbed state. The fluid flow in the matrix is assumed to be the single-phase fluid flow following Darcy's law with isotropic hydraulic conductivity. After rock failure, the observed permeability in the damage zone is generally several orders of magnitude higher than the undisturbed rock (Bossart et al. 2004). A larger permeability gives rise to quicker pore pressure dissipation in the damage zone. In the model, the damage zone permeability is increased by a factor of 100. The reasonableness of this treatment will be discussed later.

The concrete tunnel lining (15 cm in thickness) is modeled as zones with the properties of the lining material. In the model, the hydration of shotcrete is not taken into account (no time-dependent evolution of stiffness and strength) and the shotcrete is modeled as an elastic material. The Young's modulus (E_s), compressive strength (σ_s) and Poisson's ratio (v_s) for the shotcrete are E_s = 10 GPa, $\sigma_s = 10$ MPa and $v_s = 0.35$, respectively (Mont Terri Consortium 2009). The initial state of stress is assumed to be $S1$ = 6.5 MPa (vertical), $S2$ = 4.0 MPa (strike along Ga08) and $S3$ = 2.5 MPa (strike along Niche2) (Bossart and Thury 2007). These stresses are applied to the outer boundaries of the model.

It is important to note that a comparison between the model prediction results, purely based on measured parameter values as model inputs, and the experiment observations allows for a true evaluation of the predictive capability of the corresponding model within the range of parameter uncertainties that always exist in a practical application. The satisfactory matches between data and model results, obtained through model calibration (or matching the field observations through adjusting model parameters), is not necessarily a good indicator of model predictive capability beyond the range of observations used for calibration. This is because a good match can be the result of over parameterization of the model, rather than the ability of the model to capture the essence of the physical mechanisms (Li et al. 2014). Based on the above considerations, Li et al. (2014), in this case study, used the measured (rather than calibrated) parameter values given in Table 3.9 as direct inputs into the model for the mine-by test and performed model predictions of MB test results without model calibration.

3.5.2 *Simulated Results and Discussion*

Shown in Figs. 3.44a and 3.45a are the simulated damage (plastic) zones around the niche after 1.5 days, 18 days, and 25 days since the excavation. The grayscale shades distinguish between shear and tensile failure modes. The damage zone is generated by redistribution of the stress around the niche caused by mechanical unloading (excavation). Near the excavation advancing face, tensile failure is the major damage form and the maximum damage zone depth is about 2.0 m. Tensile failure is also observed in the roof and floor of the niche while shear failure occurs at the sidewalls. The damage zone penetrates about 0.7 m into the rock matrix near the roof and floor and about 1.4 m near sidewalls. Figure 3.46b shows the damage zone in a vertical cross-section at y = 12 m and perpendicular to the MB niche 2. The in situ experimental investigation revealed a dense, interconnected EDZ-fracture pattern with a maximum depth of 1.1 m from the tunnel wall; the deepest impregnated EDZ-fracture recorded was found at a depth of 2.0 m (Bossart et al. 2002, 2004; Jaeggi et al. 2012). The numerical results are generally consistent with the reported data on the EDZ. Note that the observed dense fractures within the EDZ also justify the use of the continuum modeling approach in this study (Vietor 2011).

Shown in Figs. 3.44b and 3.45b are the pore-pressure distributions around the niche after 1.5 days, 18 days and 25 days, respectively, since the excavation. The pore pressure shows a sharp change at the edge of the damage zone adjacent to the intact rock mass. Just at the niche walls, the pore pressure is 0.1 MPa that is equal

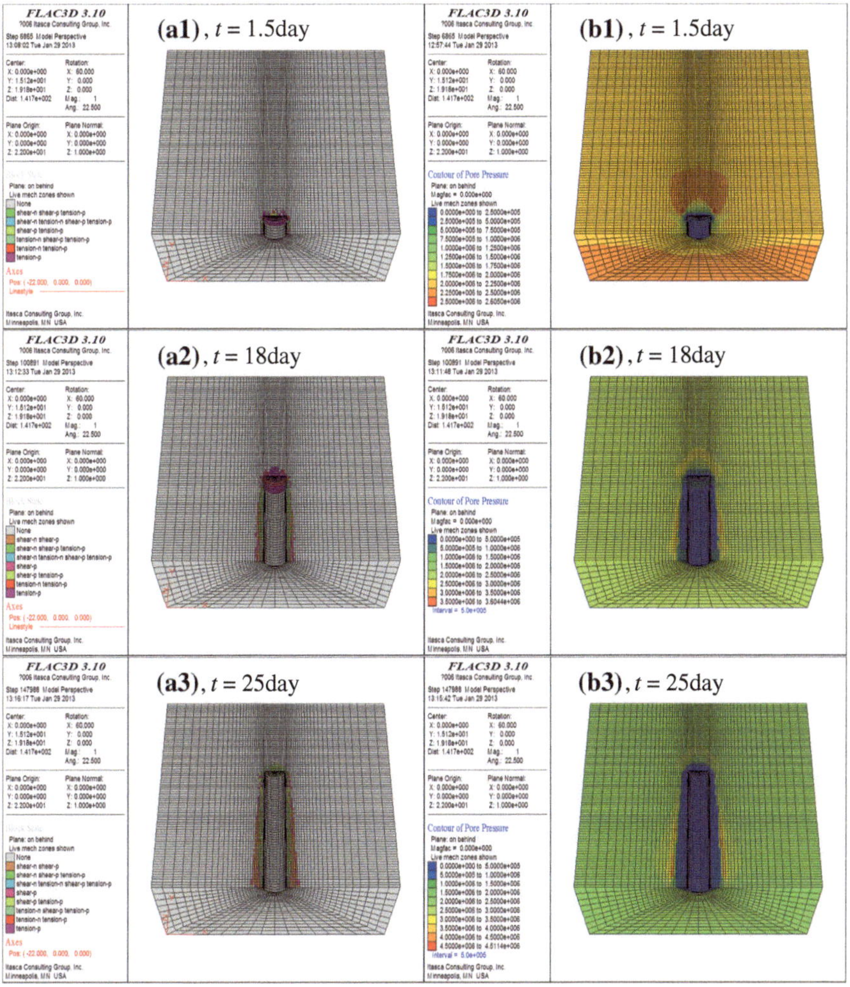

Fig. 3.44 Simulated results based on the TPHM for horizontal cross-section along MB niche 2: **a** the damage zone (and mode) and **b** pore pressure contours (Pa) (Li et al. 2014) (reproduced by permission of Elsevier)

to the atmospheric pressure because the niche is well connected to the atmosphere and thus has the atmospheric pressure. Beyond a certain distance from the niche walls in rock matrix, the pore pressure evolution shows a tendency to stabilize at a constant value of 2.0 MPa corresponding to the *in situ* pore pressure within the intact rock mass. Pore pressure increase becomes visible at about 11 m ahead of the mine-by excavation face. Near the excavation advancing face, the maximum pore water overpressure is about 3.3 MPa (as shown in Fig. 3.45b).

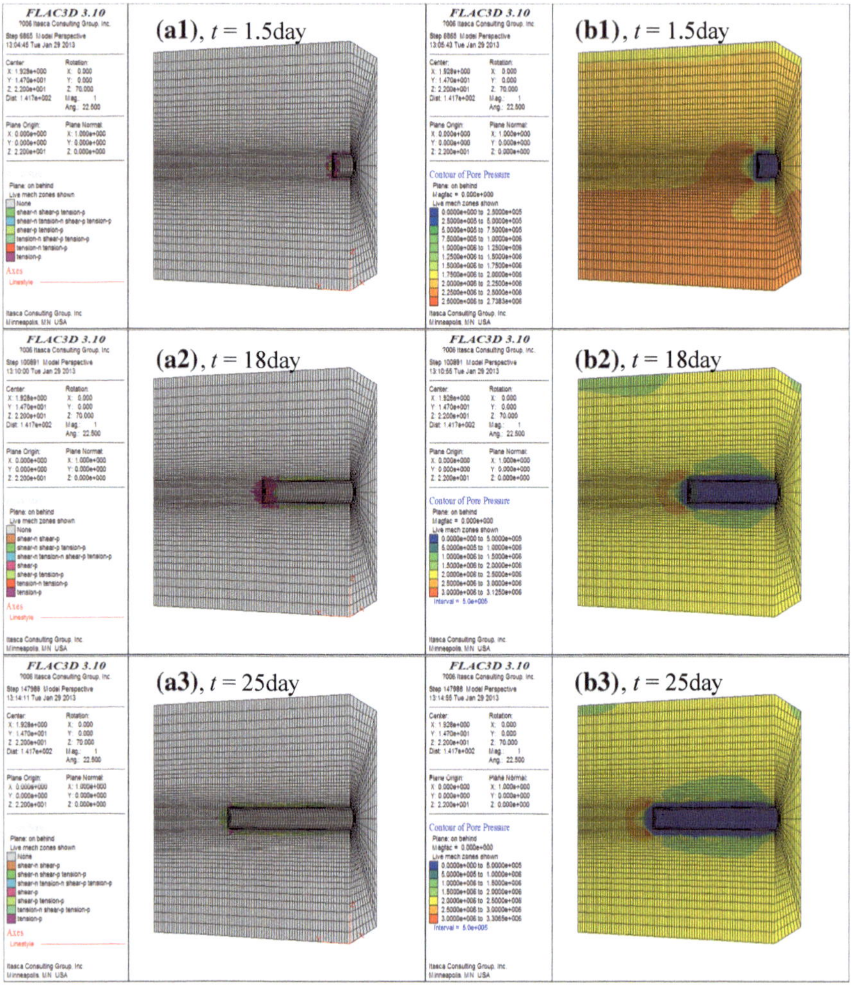

Fig. 3.45 Simulation results based on TPHM for a vertical cross-section along MB niche 2: **a** the damage zone (and mode) and **b** pore pressure contours (Pa) (Li et al. 2014) (reproduced by permission of Elsevier)

The pore pressure evolution is a relatively sensitive response to the excavation progress and there is a good correlation between the simulated damage zone and the low pore-pressure zone with pressure values similar to the niche wall pressure (0.1 MPa) (Li et al. 2014). The latter is because pore pressure within the damage zone becomes equilibrium with the niche wall pressure rapidly, as a result of that the assigned permeability for the damage zone in the model is two orders of magnitude higher than the undisturbed rock. While the observed permeability in the damage zone is several orders of magnitude higher than the undisturbed rock (Bossart et al. 2004), further increases in damage zone permeability from the two orders of magnitude do not change the simulation results significantly. Also note that air flow from niches to the damage zone likely occurred during the time that the damage zone was formed; because rock matrix permeability is very low, water flow from the matrix does not have enough time to fill in the new fractures created in the EDZ during the test. The air flow would allow the pore-pressure equilibrium between damage zone and the niche wall to take place rather quickly, given the fact that for a given permeability, the mobility of air flow is much higher than that for water flow because air has much smaller viscosity. The two-phase flow process in fractures is complex (Liu et al. 2002) and the simplification is practically needed for this model study. The good correlation between damage zone and the low pore-pressure zone also has important practical applications for monitoring EDZ evolution with pore-pressure sensors (Li et al. 2014).

Three distinguishable zones are observed from the simulated pore pressure distributions: the low-pressure zone corresponding to the damage zone, the undisturbed zone away from the damage zone, and the elevated pore-pressure zone near the edge of the damage zone (Figs. 3.44 and 3.45). The existence of the third zone is because the stress at the edge of the EDZ needs to be increased to compensate for the fact that the damage zone is not effective in supporting the overburden. The increased stress induces the elevation of pore pressure. Beyond this high pore-pressure zone into the rock, the unloading effects due to excavation and the associated damage zone gradually disappears.

As a comparison, Figs. 3.46 and 3.47 show the simulated damage zones and displacement distributions based on the TPHM and the conventional Hooke's Law or the single-part Hooke's Law (SPHM), respectively. The SPHM-based results show that the simulated damage zone develops up to an average distance of about 0.16 m into the rock near the roof and floor and up to about 0.44 m near the sidewalls. Clearly, the damage zone size from the TPHM-based model is larger than that obtained with SPHM and more consistent with field observations, as previously

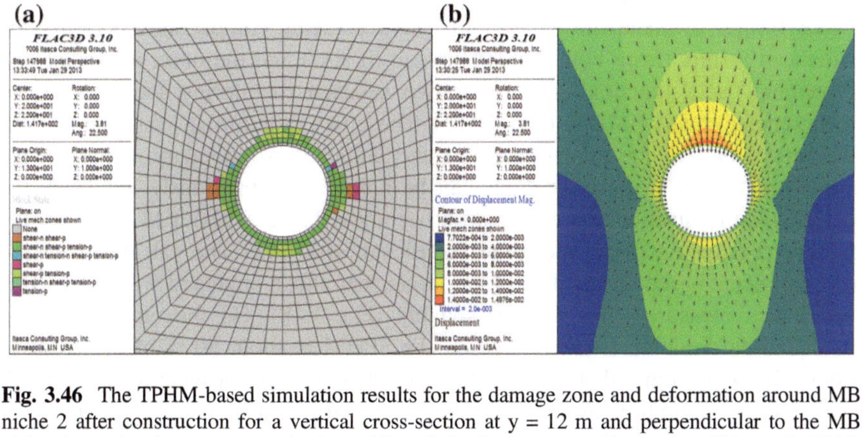

Fig. 3.46 The TPHM-based simulation results for the damage zone and deformation around MB niche 2 after construction for a vertical cross-section at y = 12 m and perpendicular to the MB niche 2: **a** the damage zone (and mode) and **b** displacement distribution (Li et al. 2014) (reproduced by permission of Elsevier)

(a) (b)

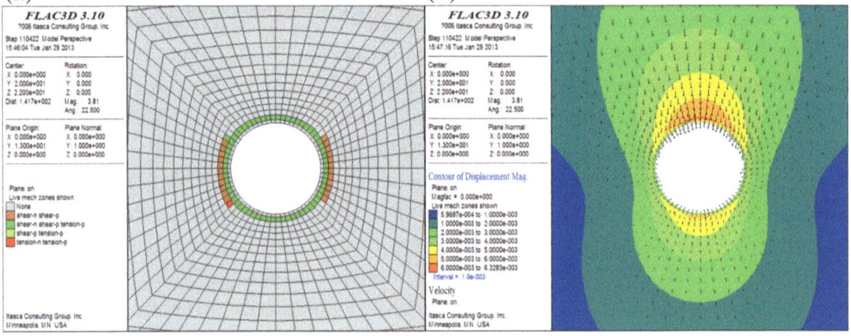

Fig. 3.47 The SPHM-based simulation results for the damage zone and deformation around MB niche 2 after construction for a vertical cross-section at Y = 12 m and perpendicular to the MB niche 2: **a** the damage zone (and mode) and **b** displacement distribution (Li et al. 2014) (reproduced by permission of Elsevier)

discussed. The better performance of the TPHM can be explained as follows for this specific problem. During the excavation, near-field unloading process occurs, resulting in a low stress range in the rock mass near the tunnel. (The normal stress on the tunnel surface is zero.) Because the "soft" part of the rock is more important in low-stress regions, rock mass surrounding the tunnel becomes much softer during excavation than it is under the ambient conditions corresponding to higher stresses. At a high stress, the "hard" part plays a more important role. This softening effect gives rise to a larger region of rock mass surrounding the tunnel (that can mechanically feel the unloading effects due to excavation) than the region simulated with the SPHM that does not consider the "soft" part. Consequently, the softening effect results in a relatively large EDZ size. The difference in the simulation results does highlight the advantages of the TPHM and the needs to use it for modeling the EDZ evolution, a critical feature determining the performance of a clay repository.

To further evaluate the TPHM, Fig. 3.48 presents comparisons between the simulated results (from both the TPHM and SPHM) and the observed radial displacement, as a function of time, at two selected monitoring locations at the inclinometer chains BMB30 (Fig. 3.48c). These two monitoring locations are near the central part of the model such that the influence of boundary effects on the numerical results at these locations can be ignored. As previously indicated, the deformation of surrounding rock matrix becomes visible at about 11 m before the mine-by excavation front and more significant with time. The calculated inward displacements based on the TPHM are larger than those predicted with the SPHM and match measured data very reasonably (Fig. 3.48). On the other hand, the predicted deformation (strain) based on the SPHM is about 50 % of the observations. One may argue that the SPHM-based model could provide a better match with model calibration by adjusting values for the related parameters. It is the author's belief that model calibration makes sense only when the model contains the correct physical mechanisms. Otherwise, the good match to observations would

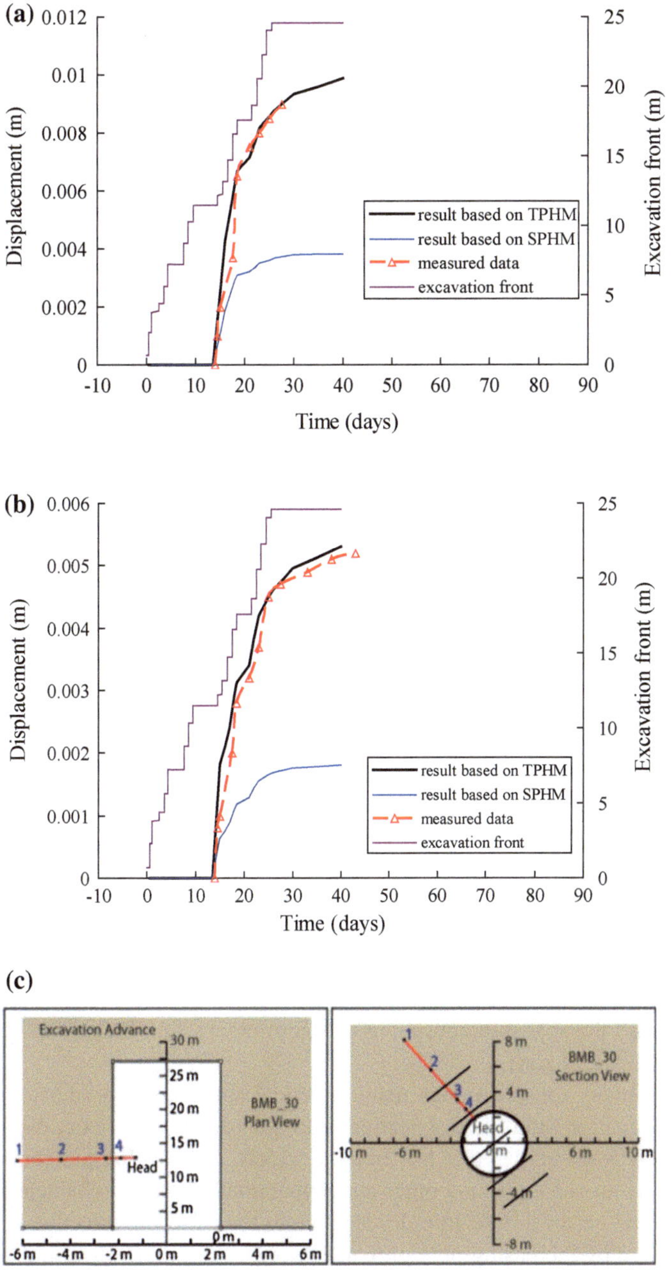

Fig. 3.48 Excavation advancing and (simulated and observed) inward displacements as a function of time (Li et al. 2014) (reproduced by permission of Elsevier). **a** Displacement at sensor BMB30_4. **b** Displacement at sensor BMB30_1. **c** The locations of displacement monitoring sensors (BMB30)

not be very meaningful and could not be used as an indicator of the model's predictive capability either, as previously indicated. Nevertheless, the better performance of the TPHM-based model is a result of the fact that the TPHM takes into account the mechanical behavior of "soft" features within the rock mass.

Figure 3.49 shows comparisons between the simulated and the observed pore pressures (as functions of time) at two locations corresponding to sensors BMB15 and BMB16 (Fig. 3.49c) that are 16.0 and 22.0 m, respectively, away from the beginning of the niche along the niche's longitudinal direction. The observations can be explained based on the evolution of the three pore-pressure zones discussed above. The sensors were initially located within the undisturbed rock mass and therefore provided flat pore pressure distributions with a pressure value corresponding to the ambient condition. With the excavation advancing, the locations became closer to the evolving EDZ and reached the elevated pressure zone where the stress (and pore pressure) at the edge of EDZ needs to be increased to compensate for the fact that the damage zone is not effective in supporting the overburden. Then when the sensor locations were within the EDZ, the pore pressures started to decline rapidly with time because the EDZ has a relatively large permeability that allows for quick pore-pressure dissipation. The observed pore-pressure histories at two sensor locations are realistically predicted by the TPHM simulation results. The SPHM does a good job as well for sensor BMB16 (Fig. 3.49b), but considerably underestimates pore pressure decline for BMB15, because the SPHM underestimates the EDZ size and consequently underestimates the pore-pressure dissipation rate at the sensor (BMB 15) location.

Given the fact that Li et al. (2014), in this case study, directly used the measured property data as model inputs and did not conduct any model calibrations, it is remarkable that the TPHM-based simulation results have excellent agreements with observations of different kinds from the MB tests, including the sizes of damage zone, displacements near the niche, and pore pressure evolution at multiple locations. While results from a single case-study should be interpreted with caution, this study does clearly indicate that the model results from the conventional Hooke's Law (or the SPHM) can significantly underestimate the mechanical deformation and the damage near a tunnel because the SPHM is not able to capture the contribution from the soft part consisting of micro-cracks that becomes significant in low effective stress zones.

The low stress zones occur in many practical applications associated with subsurface excavation and fluid injection into the depth. For example, hydraulic fracturing of horizontal wells has been a widely used technique for recovering natural gas from shale gas and other unconventional reservoirs characterized by an extremely low permeability. In order to fracture the rock, the effective stress needs to be negative by increasing pore pressure through fluid injection (Eq. 3.68). Thus, there must be rock zones near the generated hydraulic fractures that should have low effective stress if the injected fluid can flow into these zones to increase pore pressure there during the hydraulic fracturing process. In the other words, micro-cracks, or the soft part, in these zones are stimulated. Because these cracks may dominate rock-matrix permeability for low-permeability media, stimulation of

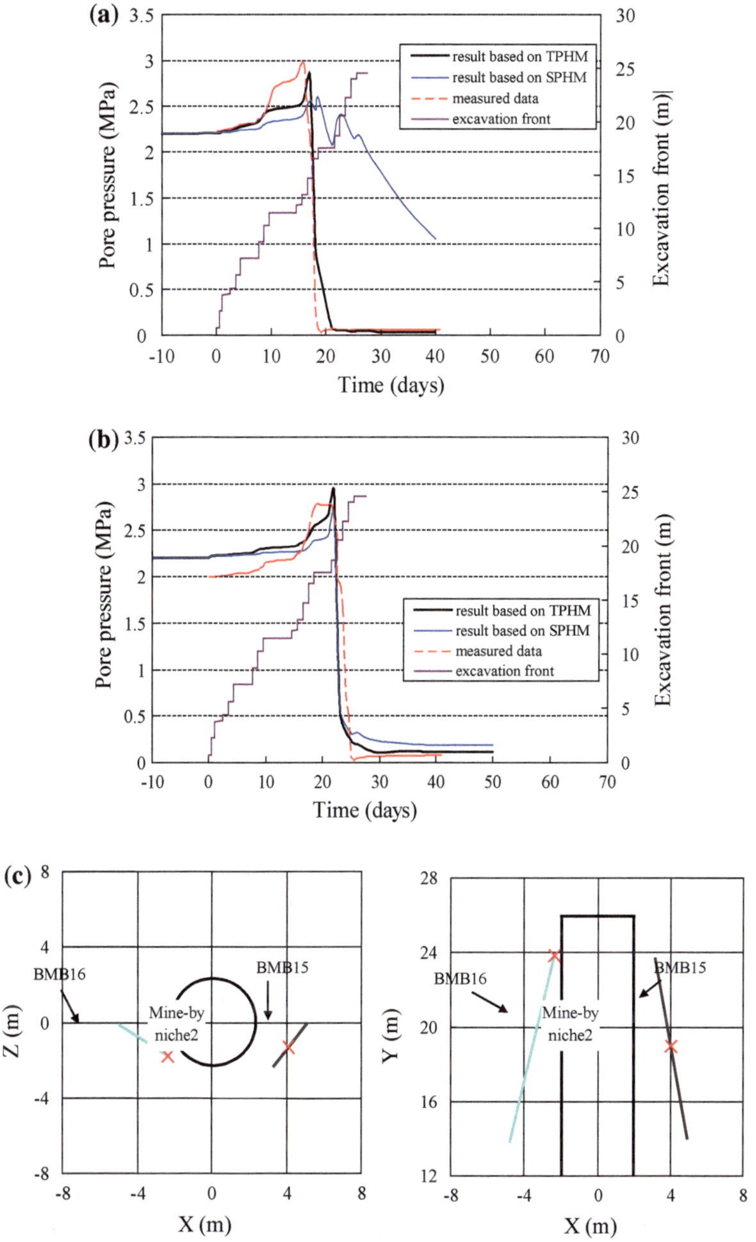

Fig. 3.49 Excavation advancing and (simulated and observed) pore pressures as a function of time (Li et al. 2014) (reproduced by permission of Elsevier). **a** Pore pressure at sensor BMB15. **b** Pore pressure at sensor BMB16. **c** The locations of pore-pressure monitoring sensors (BMB15 and BMB16)

these cracks is critical for recovering hydrocarbon fluid flow from the rock matrix in unconventional reservoirs. The TPHM is useful for modeling the stimulation of these micro-cracks and their impact on hydraulic and mechanical properties.

3.6 Concluding Remarks

(A) The key results from this chapter are the TPHM given by Eq. 3.8 and its variations:

$$-\frac{dV}{V_0} = \gamma_e \frac{d\sigma}{K_e} + \gamma_t \exp\left(-\frac{\sigma}{K_t}\right)\frac{d\sigma}{K_t} \tag{3.8}$$

where V is the rock volume, V_0 is the rock volume under zero stress, σ is the stress, γ_e and γ_t are the volumetric fractions of hard and soft parts of a rock, respectively, under zero stress and K_e and K_t refer to bulk moduli for the hard and soft parts, respectively. Without the soft part, the TPHM is reduced to conventional Hooke's Law. The development is based on a well-known fact that micro-cracks (soft part) exist in natural rock and exhibit very different mechanical behavior from the rest of rock mass. Note that micro-cracks can serve as the dominant flow pathways with a rock matrix in many cases.

(B) The generality of TPHM is demonstrated in the chapter. It allows for derivation of a variety of constitutive relationships. We show the consistence of these relationships with empirical relationships and data sets from different sources. Furthermore, the TPHM is shown to be applicable to both rock matrix and fractures.

(C) The TPHM needs to be further developed at least in two aspects. Firstly, the TPHM has been developed for mechanically isotropic medium for both isotropic and anisotropic stress conditions. It needs to be extended to anisotropic media (such as bedding media). Secondly, the current version of the TPHM is limited to elastic deformations. How to relate the TPHM to micro-crack evolution during the damage process should be investigated.

References

Al-Tahini A, Abousleiman Y (2010) Pore-pressure-coefficient anisotropy measurements for intrinsic and induced anisotropy in sandstone. SPE Reservior Eval Eng, 265–274

Armero F (1999) Formulation and finite element implementation of a multiplicative model of coupled poro-plasticity at finite strains under fully saturated conditions. Comput Methods Appl Mech Eng 171:205–241

Armero F, Simo JC (1992) A new unconditionally stable fractional step method for nonlinear coupled thermomechanical problems. Int J Numer Methods Eng 35:737–766

Athy LF (1930) Density, porosity, and compaction of sedimentary rock. Am Assoc Pet Geol Bull 14:1–24

Bai M, Elsworth D, Roegiers JC (1993) Multiporosity/multipermeability approach to the simulation of naturally fractured reservoirs. Water Resour Res 29:1621–1633

Bandis SC, Lumsden AC, Barton NR (1983) Fundamentals of rock joint deformation. Int J Rock Mech Min Sci Geomech Abstr 20(6):249–268

Barenblatt GI, Zheltov YP (1960) Fundamental equations of filtration of homogeneous liquids in fissured rocks. Sov Phys Dokl Engl Transl 5:522–525

Barton N, Bandis S, Bakhtar K (1985) Strength, deformation and conductivity coupling of rock joints. Int J Rock Mech Min Sci Geomech Abstr 22:121–140

Becker K, Shapiro SA, Stanchits S et al (2007) Stress induced elastic anisotropy of the Etnean basalt: theoretical and laboratory examination. Geophy Res Lett 34:L11307

Berryman JG (1992) Effective stress for transport properties of inhomogeneous porous rock. J Geophys Res 97(B12):17409–17424

Berryman JG (2006) Estimates and rigorous bounds on pore-fluid enhanced shear modulus in poroelastic media with hard and soft anisotropy. Int J Damage 15(2):133–167

Berryman JG, Wang HF (1995) The elastic coefficients of double-porosity models for fluid transport in jointed rock. J Geophys Res 100(B12):24611–24627

Biot MA (1941) General theory of three dimensional consolidation. J Appl Phys 12:155–164

Biot MA (1956) General solutions of the equations of elasticity and consolidation for a porous material. J Appl Mech Trans ASME 78:91–96

Blumling P, Bernier F, Lebon P et al (2007) The excavation damaged zone in clay formations time-dependent behavior and influence on performance assessment. Phys Chem Earth 32:588–599

Bock H (2009) RA experiment: updated review of the rock mechanics properties of the opalinus clay of the Mont Terri URL based on laboratory and field testing. Technical report 2008-04, Q+S Consult, Germany

Bock H, Dehandschutter B, Martin CD (2010) Self-sealing of fractures in argillaceous formations in the context of geological disposal of radioactive waste, review and synthesis. Nuclear Energy Agency, ISBN 978-92-64-99095-1

Bossart P, Thury M (2007) Mont Terri Rock Laboratory—Project, Programme 1996 to 2007 and Results. Report No. 3—Swiss Geological Survey, Wabern

Bossart P, Meier PM, Moeri A et al (2002) Geological and hydraulic characterization of the excavation disturbed zone in the Opalinus Clay of the Mont Terri Rock Laboratory. Eng Geol 66:19–38

Bossart P, Meier PM, Moeri A et al (2004) Structural and hydrogeological characterisation of the excavation-disturbed zone in the Opalinus Clay (Mont Terri Project, Switzerland). Appl Clay Sci 26:429–448

Brighenti G (1989) Effect of confining pressure on gas permeability of tight sandstones. ISRM international symposium, International Society for Rock Mechanics

Brooks RH, Corey AT (1964) Hydraulic properties of porous media. Hydrology paper no. 3, Civil Engineering Dep., Colorado State Univ., Colorado, USA

Brown SR (1987) Fluid flow through rock joints: the effect of surface roughness. J Geophys Res Solid Earth 92:1337–1347

Brown SR, Scholz CH (1985) Closure of random surfaces in contact. J Geophys Res 90 (B7):5531–5545

Brown SR, Scholz CH (1986) Closure of rock joints. J Geophys Res 91(B5):4939–4948

Burdine NT (1953) Relative permeability calculations from pore-size distribution data. Petr Trans Am Inst Mining Metall Eng 198:71–77

Byrnes AP (1997) Reservoir characteristics of low-permeability sandstones in the Rocky Mountains. The Mountain Geologist

Byrnes AP, Castle JW (2000) Comparison of core petrophysical properties between low-permeability sandstone reservoirs: Eastern US Medina group and Western US Mesaverde group and Frontier formation. Paper presented in SPE Rocky Mountain Regional/low-permeability reservoirs symposium and exhibition, Society of Petroleum Engineers

Carpenter CB, Spencer GB (1940) Measurement of compressibility of consolidated oil-bearing sandstones. Report 3540, US Bureau of Mines, Denver, CO

Chen CY, Horne RN (2006) Two-phase flow in rough-walled fractures: experiments and a flow structure model. Water Resour Res. doi:10.1029/2004WR003837

Chen Z, Narayan SP, Yang Z et al (2000) An experimental investigation of hydraulic behavior of fractures and joints in granitic rock. Int J Rock Mech Min Sci 37:1061–1071

Corkum AG, Martin CD (2007a) The mechanical behavior of weak mudstone (Opalinus Clay) at low stresses. Int J Rock Mech Min Sci 21:196–209

Corkum AG, Martin CD (2007b) Modeling a mine-by test at the Mont Terri Rock Laboratory, Switzerland. Int J Rock Mech Min Sci 44:846–859

Coyner KB (1984) Effects of stress, pore pressure, and pore fluids on bulk strain, velocity, and permeability of rocks. Ph. D. thesis. Mass Inst Technol, Cambridge, MA

Crease RP (2010) The great equations: breakthroughs in science from Pythagoras to Heisenberg. W W Norton & Company, New York

Cui X, Bustin RM, Chikatamarla L (2007) Adsorption-induced coal swelling and stress: implications for methane production and acid gas sequestration into coal seams. J Geophys Res. doi:10.1029/2004JB003482

David C, Wong TF, Zhu W (1994) Laboratory measurement of compaction-induced permeability change in porous rocks: implications for the generation and maintenance of pore pressure excess in the crust. Pure Appl Geophys 143(1–3):425–456

David C, Menendez B, Zhu W et al (2001) Mechanical compaction, microstructures and permeability evolution in sandstones. Phys Chem Earth Part A 26(1):45–51

Davis JP, Davis DK (1999) Stress-dependent permeability: characterization and modeling. SPE paper no. 56813, Society of Petroleum Engineers

Davy CA, Skoczylas F, Barnichon JD et al (2007) Permeability of macro-cracked argillite under confinement: gas and water testing. Phys Chem Earth 32:667–680

Dewhurst DN, Aplin AC, Sarda JP et al (1998) Compaction-driven evolution of porosity and permeability in natural mudstones: an experimental study. J Geophys Res Solid Earth 103 (B1):651–661

Diomampo GP (2001) Relative permeability through fractures. Rep. SGP-TR-170, Stanford University, CA

Dong JJ, Hsu JY, Wu WJ (2010) Stress-dependence of the permeability and porosity of sandstone and shale from TCDP Hole-A. Int J Rock Mech Min Sci 47(7):1141–1157

Durham WB, Bonner BP (1994) Self-propping and fluid flow in slightly offset joints at high effective pressures. J Geophys Res 99(B5):9391–9399

Elliot GM, Brown ET (1988) Laboratory measurement of the thermo-hydro-mechanical properties of rock. Q J Eng Geol Hydrogeol 21:299–314

Evans JP, Forster CB, Goddard JV (1997) Permeability of fault-related rocks, and implications for hydraulic structure of fault zones. J Struct Geol 19(11):1393–1404

Fjær E, Holt RM, Horsrud P et al (2008) Petroleum related rock mechanics, 2nd edn. Elsevier, Boston

Fourar M, Lenormand R (1998) A viscous coupling model for relative permeabilities in fractures. Paper SPE 49006 presented at the SPE annual technical conference and exhibition, New Orleans, Louisiana, USA

Freed AD (1995) Natural strain. J Eng Mater Technol 117:379–385

Gangi AF (1978) Variation of whole and fractured porous rock permeability with confining pressure. Int J Rock Mech Min Sci Geomech Abstr 15:249–257

Gercek H (2007) Poisson's ratio values for rocks. Int J Rock Mech Min Sci 44:1–13

Ghabezloo S, Sulem J, Martineau F (2009a) Effective stress law for the permeability of a limestone. Int J Rock Mech Min Sci 46(2):297–306

Ghabezloo S, Sulem J, Saint-Marc J (2009b) Evaluation of a permeability-porosity relationship in a low-permeability creeping material using a single transient test. Int J Rock Mech Min Sci 46 (4):761–768

Goodman RE (1974) The mechanical properties of joints. In Proceedings 3rd Congress ISRM, Denver, vol 1, pp 127–140

Goodman RE (1976) Method of geological engineering in discontinuous rocks. West Publishing, New York

Harpalani S, Zhao X (1989) An investigation of the effect of gas desorption on gas permeability. In: Proceedings of the coalbed methane symposium, University of Alabama, Tuscaloosa, Alabama, pp 57–64

Hu Y, Liu GT (2004) Behavior of soft rock under multiaxial compression and its effects on design of arch dam. Chin J Rock Mech Eng 23:2494–2498

Indraratna B, Ranjith PG, Gale W (1999) Single phase water flow through rock fractures. Geotech Geol Eng 17:211–240

Indraratna B, Ranjith PG, Gale W (2002) Some aspects of unsaturated flow in jointed rock. Int J Rock Mech Min Sci 39:555–568

Inwood S (2005) The forgotten genious: the biography of Robert Hooke. MacAdam/Cage Publishing

Itasca Consulting Group (2005) FLAC3D: fast lagrangian analysis of continuain 3dimensions. Itasca Consulting Group, Minneapolis

Iwai K (1976) Fundamental studies of fluid flow through a single fracture. Ph. D. thesis, University of California, Berkeley, CA

Iwano M (1995) Hydromechanical characteristics of a single rock joint. Ph. D. thesis, Massachusetts Institute of Technology, Cambridge, MA

Jaeger JC, Cook NGW, Zimmerman RW (2007) Fundamentals of rock mechanics, 4th edn. Blackwell Publishing, Malden

Jaeggi D, Nussbaum C, Vietor T (2012) MB experiment: visualization of EDZ-fractures around the MB niche by using the in-situ resin impregnation technique. Mont Terri Project TN 2010-56

Jardine L (2004) The curious life of Robert Hooke: the man who measured London. Harper Collins Publishers, New York

Jasinge D, Ranjith PG, Choi SK (2011) Effects of effective stress changes on permeability of latrobe valley brown coal. Fuel 90(3):1292–1300

Jobmann M, Wilsnack Th, Voigt HD (2010) Investigation of damage-induced permeability of Opalinus Clay. Int J Rock Mech Min Sci 46:279–285

Jones FO, Owens WW (1980) A laboratory study of low-permeability gas sands. J Petrol Technol 32(9):1631–1640

Kazemi H, Gilman JR (1993) Multiphase flow in fractured petroleum reservoirs. In: Bear J, Tsang CF, de Marsily G (eds) Flow and transport in fractured rock. Academic Press, Berkeley, CA

Kilmer NH, Morrow NR, Pitman JK (1987) Pressure sensitivity of low permeability sandstones. J Petrol Sci Eng 1(1):65–81

Konecny P, Kozusnikova A (2011) Influence of stress on the permeability of coal and sedimentary rocks of the Upper Silesian basin. Int J Rock Mech Min Sci 48(2):347–352

Kwon O, Kronenberg AK, Gangi AF et al (2012) Permeability of Wilcox shale and its effective pressure law. J Geophys Res Solid Earth 106(B9):19339–19353

Lei Q, Xiong W, Yuan J et al (2007) Analysis of stress sensitivity and its influence on oil production from tight reservoirs. Paper SPE 111148 presented at the 2007 SPE Eastern Regional meeting held in Lexington, Kentucky, USA, 17–19 Oct 2007

Li L, Liu HH, Birkholzer J et al (2014) The use of two-part Hooke's model (TPHM) to model the mine-by test at Mont Terri Site, Switzerland. Comput Geotech 58:28–46

Lin W, Tang GQ, Kovscek AR (2008) Sorption-induced permeability changes of coal during gas-injection processes. SPE Reserv Eval Eng 11(4):792–802

Liu HH, Rutqvist J (2010) A new coal-permeability model: internal swelling stress and fracture–matrix interaction. Transp Porous Media 82:157–171

Liu HH, Rutqvist J (2013) Coupled hydro-mechanical processes associated with multiphase flow in a dual-continuum system: formulations and an application. Rock Mech Rock Eng 46 (5):1103–1112

Liu HH, Doughty C, Bodvarsson GS (1998) An active fracture model for unsaturated flow and transport in fractured rocks. Water Resour Res 34:2633–2646

Liu HH, Bodvarsson GS, Finsterle S (2002) A note on unsaturated flow in two-dimensional fracture networks. Water Resour Res 38(9):15-1–15-9

Liu HH, Rutqvist J, Berryman JG (2009) On the relationship between stress and elastic strain for porous and fractured rock. Int J Rock Mech Min Sci 46:289–296

Liu HH, Rutqvist J, Birkholzer JT (2011) Constitutive relationships for elastic deformation of clay rock: data analysis. Rock Mech Rock Eng. doi:10.1007/s00603-010-0131-4

Liu HH, Wei MY, Rutqvist J (2013) Normal-stress dependence of fracture hydraulic properties including two-phase flow properties. Hydrogeol J 21(2):371–382

MacBeth C (2004) A classification for the pressure-sensitivity properties of a sandstone rock frame. Geophysics 69:497–510

Mainguy M, Longuemare P (2002) Coupling fluid flow and rock mechanics: formulations of partial coupling between reservoir and geomechanical simulators. Oil Gas Sci Technol 57:355–367

Malama B, Kulatilake PHSW (2003) Models for normal fracture deformation under compressive loading. Int J Rock Mech Min Sci 40:893–901

Matsuki K, Wang EQ, Sakaguchi K et al (2001) Time-dependent closure of a fracture with rough surfaces under constant normal stress. Int J Rock Mech Min Sci 38:607–619

Mavko G, Jizba D (1991) Estimating grain-scale fluid effects on velocity dispersion in rocks. Geophysics 56:1940–1949

Mavko G, Nur A (1978) Melt squirt in the asthenosphere. J Geophys Res 80:1444–1448

Mavor MJ, Vaughn JE (1997) Increasing absolute permeability in the San Juan basin Fruitland formation. In: Proceedings of the coalbed methane symposium, University of Alabama, Tuscaloosa, AL

McDermott CI, Kolditz O (2006) Geomechanical model for fracture deformation under hydraulic, mechanical and thermal loads. Hydrogeol J 14(4):487–498

McLatchie A, Hemstock RA, Young JW (1958) The effective compressibility of reservoir rock and its effects on permeability. J Petrol Technol 10(06):49–51

Metwally YM, Sondergeld CH (2011) Measuring low permeabilities of gas-sands and shales using a pressure transmission technique. Int J Rock Mech Min Sci 48(7):1135–1144

Mohiuddin M, Kovin G, Abdulraheem A et al (2000) Stress-dependent porosity and permeability of a suite of samples from Saudi Arabian sandstone and limestone reservoirs. Paper SCA2033 presented in symposium of core analysts, Abu Dhabi, UAE

Mont Terri Consortium (2009) Characterisation of the dry shotcrete sprayed in MB niche for modelling works. Final report, Geotechnisches Institut AG, Bern

Morgenstern NR, Tamuly Phukan AL (1969) Nonlinear stress-strain relations for a homogeneous sandstone. Int J Rock Mech Min Sci 6:127–142

Moyer AE (1977) Robert Hooke ambiguous presentation of “Hooke’s Law”. Isis 68(2):266–275

Mualem Y (1976) A new model of predicting the hydraulic conductivity of unsaturated porous media. Water Resour Res 12:513–522

Nagra (2002) Projekt Opalinuston – Synthese der geowissenschaftlichen Untersuchungsergebnisse. Entsorgungsnachweis für abgebrannte Brennelemente, verglaste hochaktive sowie langlebige mittelaktive Abfälle. Nagra technical report NTB 02-03, Nagra, Wettingen, Switzerland

National Research Council (1996) Rock fractures and fluid flow: contemporary understanding and applications. National Academy Press, Washington DC

Neuzil CE (2003) Hydromechanical coupling in geologic processes. Hydrogeol J 11:41–83
Nur A (1971) Effects of stress on velocity anisotropy in rocks with cracks. J Geophys Res 76:2022–2034
Olalla C, Martin M, Sáez J (1999) ED-B experiment: geotechnical laboratory test on Opalinus Clay rock samples. Technical report TN98-57, Mont Terri Project
Olivella S, Carrera J, Gens A et al (1994) Nonisothermal multiphase flow of brine and gas through saline media. Transp Porous Media 15:271–293
Olsson R, Barton N (2001) An improved model for hydromechanical coupling during shearing of rock joints. Int J Rock Mech Min Sci 39:555–568
Olsson WA, Brown SR (1993) Hydromechanical response of a fracture undergoing compression and shear. Int J Rock Mech Min Sci Geomech Abstr 30:845–851
Oron AP, Berkowitz B (1998) Flow in rock fractures: the local cubic law assumption reexamined. Water Resour Res 34:2811–2825
Palmer I, Mansoori J (1998) How permeability depends on stress and pore pressure in coalbeds: a new model. SPE Reserv Eval Eng 1(6):539–544
Peng SP, Zhang JC (2007) Engineering geology for underground rocks. Springer, Berlin
Poisson SD (1829) Mémoire sur les equations generates de l'équilibre et du nouvement des corps silides élastiques et de fluids. Journal de l'École Poly technique 13:1–174
Popp T, Salzer K, Minkley W (2008) Influence of bedding planes to EDZ-evolution and the coupled HM properties of Opalinus clay. Phys Chem Earth 33:5374–5387
Poulos HG, Davis EH (1974) Elastic solutions for soil and rock mechanics. Wiley, New York
Power WL, Tullis TE (1992) Euclidean and fractal models for the description of rock surface roughness. J Geophys Res Solid Earth 96:415–424
Pruess K (1991) TOUGH2—a general purpose numerical simulator for multiphase fluid and heat flow. Lawrence Berkeley National Laboratory report LBL-29400
Pruess K, Tsang YW (1990) On two-phase relative permeability and capillary pressure of rough-walled rock fractures. Water Resour Res 26(9):1915–1926
Pyrak-Nolte LJ, Myer LR, Cook NGW et al (1987) Hydraulic and mechanical properties of natural fractures in low-permeability rock. In: 6th ISRM congress, Montreal, Canada
Raven KG, Gale JE (1985) Water flow in a natural rock fracture as a function of stress and sample size. Int J Rock Mech Min Sci Geomech Abstr 22(4):251–261
Robertson EP (2005) Measurement and modeling of sorption-induced strain and permeability changes in coal. Report INL/EXT-06-11832, Idaho National Laboratory
Rutqvist J, Borgesson L, Chijimatsu M et al (2001) Thermohydromechanics of partially saturated geological media: governing equations and formulation of four finite element models. Int J Rock Mech Min Sci 38:105–127
Rutqvist J, Wu YS, Tsang CF et al (2002) A modeling approach for analysis of coupled multiphase fluid flow, heat transfer, and deformation in fractured porous rock. Int J Rock Mech Min Sci 39:429–442
Sayers CM, Kachanov M (1995) Microcrack-induced elastic wave anisotropy of brittle rocks. J Geophys Res 100:4149–4156
Schoenberg M, Sayers CM (1995) Seismic anisotropy of fractured rock. Geophysics 60:204–211
Schrauf TW, Evans DD (1986) Laboratory studies of gas flow through a single natural fracture. Water Resour Res 22(7):1038–1050
Settari A, Mourits FM (1998) A coupled reservoir and geomechanical simulation system. Soc Pet Eng J 3(3):219–226
Settari A, Walters DA (2001) Advances in coupled geomechanical and reservoir modeling with applications to reservoir compaction. Soc Pet Eng J 6(3):334–342
Shapiro SA (2003) Piezosensitivity of porous and fractured rocks. Geophysics 68:482–486
Shapiro SA, Kaslow A (2005) Porosity and elastic anisotropy of rocks under tectonic stress and pore pressure changes. Geophysics 70:N27–N38
Sharifzadeh M, Mitani Y, Esaki T (2008) Rock joint surface measurement and analysis of aperture distribution under different normal and shear loading using GIS. Rock Mech Rock Eng 41 (2):299–323

Shi Y, Wang CY (1986) Pore pressure generation in sedimentary basins: overloading versus aquathermal. J Geophys Res Solid Earth 91(B2):2153–2162
Spencer CW (1989) Review of characteristics of low-permeability gas reservoirs in western United States. AAPG Bull 73(5):613–629
Therrien R, Sudicky EA (1996) Three-dimensional analysis of variably-saturated flow and solute transport in discretely-fractured porous media. J Contam Hydrol 23:1–44
Thomas RD, Ward DC (1972) Effect of overburden pressure and water saturation on gas permeability of tight sandstone cores. J Petrol Technol 24(02):120–124
Tsang CF, Barnichon JD, Birkholzer J et al (2012) Coupled thermo-hydro-mechanical processes in the near field of a high-level radioactive waste repository in clay formations. Int J Rock Mech Min Sci 49:31–44
Vairogs J, Hearn CL, Dareing DW et al (1971) Effect of rock stress on gas production from low-permeability reservoirs. J Petrol Technol 23(09):1161–1167
Vietor T (2011) Mine-by test in Mont Terri rock laboratory-MB experiment report: data base for the predictive model. Nagra, Switzerland
Walls JD, Nur AM, Bourbie T (1982) Effects of pressure and partial water saturation on gas permeability in tight sands: experimental results. J Petrol Technol 34(04):930–936
Walsh JB (1965a) The effect of cracks on the compressibility of rock. J Geophys Res 70:381–389
Walsh JB (1965b) The effect of cracks on the uniaxial elastic compression of rocks. J Geophys Res 70:399–411
Walsh R, McDermott C, Kolditz O (2008) Numerical modeling of stress-permeability coupling in rough fractures. Hydrogeol J 16:613–627
Warren JE, Root PJ (1963) The behavior of naturally fractured reservoirs. Soc Pet Eng J 3:245–255
Weerakone WMSB, Wong RCK, Mehrotra AK (2011) Single-phase (Brine) and two-phase (DNAPL-Brine) flow in induced fractures. Transp Porous Med 89:75–95
Wei MY, Liu HH, Li LC et al (2013) A fractal-based model for fracture deformation under shearing and compression. Rock Mech Rock Eng 46(6):1539–1549
Wheeler MF, Gai X (2007) Iteratively coupled mixed and Galerkin finite element methods for poro-clasticity. Numer Meth PDEs 23:785–797
White CM, Smith DH, Jones KL et al (2005) Sequestration of carbon dioxide in coal with enhanced coalbed methane recovery—a review. Energy Fuels 19:559–724
Wilson RK, Aifantis EC (1982) On the theory of consolidation with double porosity. Int J Eng Sci 20:1009–1035
Witherspoon PA, Wang JSW, Iwai K et al (1980) Validity of cubic law for fluid flow in deformable rock fracture. Water Resour Res 16(6):1016–1024
Wong RCK, Pan X, Maini BB (2008) Correlation between pressure gradient and phase saturation for oil-water flow in smooth- and rough-walled parallel plate models. Water Resour Res. doi:10.1029/2007/WR006043
Wright I (2008) Satellite monitoring at In Salah. Seventh annual conference on carbon capture and sequestration, Pittsburg, Pennsylvania, 5–8 May
Wyble DO (1958) Effect of applied pressure on the conductivity, porosity, and permeability of sandstones. Petrol Trans AIME 213:430–432
Xu GM, Liu QS, Peng WW (2006) Experimental study on basic mechanical behaviors of rocks under low temperatures. Chin J Rock Mech Eng 25:2502–2508
Yale DP, Nur A (1985) Network modeling of flow storage and deformation in porous rocks. SEG annual meeting, Society of Exploration Geophysicists
Zhang CL, Rothfuchs T (2008) Damage and sealing of clay rocks detected by measurements of gas permeability. Phys Chem Earth 33:5363–5373
Zhang CL, Rothfuchs T, Dittrich J et al (2008) Investigations on self-sealing of indurated clay. GRS report GRS-230:1-67, ISBN 978-3-939355-04-5
Zhao J, Brown ET (1992) Hydro-thermo-mechanical properties of joints in the Carnmenellis granite. Q J Eng Geol 25:279–290

Zhao Y, Liu HH (2012) An elastic strain-stress relationship for porous rock under anisotropic stress conditions. Rock Mech Rock Eng. doi:10.1007/s00603-011-0193-y

Zheng JT, Zheng LG, Liu HH et al (2015) Relationships between permeability, porosity and effective stress for low-permeability sedimentary rock. Int J Rock Mech Min Sci 78:304–318

Zienkiewicz OC, Paul DK, Chan AHC (1988) Unconditionally stable staggered solution procedure for soil-pore fluid interaction problems. Int J Numer Methods Eng 26(5):1039–1055

Zimmerman RW (1991) Compressibility of sandstones. Elsevier, Amsterdam

Zimmerman RW, Bodvarsson GS (1996) Hydraulic conductivity of rock fractures. Transp Porous Media 23(1):1–30

Chapter 4
A Thermodynamic Hypothesis Regarding Optimality Principles for Flow Processes in Geosystems

Abstract While optimality principles have been successfully used in many different areas related to flow processes in geosystems, their thermodynamic base has not been fully established. As an attempt to address this important issue, this chapter presents a thermodynamic hypothesis regarding optimality principles for the flow process. It states that *a nonlinear natural system that is not isolated and involves positive feedbacks tends to minimize its resistance to the flow process that is imposed by its environment.* Consistence between the hypothesis and typical flow processes in geosystems is demonstrated. In spirit, the hypothesis is consistent with Darwin's evolution theory. It reconciles the seeming inconsistency between the minimization of energy expenditure rate principle and the maximum entropy production principle. An application of the hypothesis to calculation of inelastic deformation of natural rock is also outlined. The hypothesis is fundamental in nature, but proposed in a phenomenological manner. Further examinations of the usefulness and potential limitations of the hypothesis in describing other processes, distinct from flow processes in geosystems, are needed.

An optimality principle refers to that state of a physical system is controlled by an optimal condition being subject to physical and resource constraints. As discussed in Chap. 2, the principles of this kind have a long history in science, and they have been proven to be useful for many engineering applications. It is, however, surprising that the thermodynamic basis for these principles has not been fully established yet, probably because the non-equilibrium thermodynamics is still a scientific area under active development (Glansdorff and Prigogine 1978; Eigen 2013). This short chapter focuses on a newly proposed thermodynamic hypothesis for flow process in geosystems, mainly based on the materials from Liu (2014). Note that the definitions and physical meanings of the same symbol (denoting a variable or function) herein may be different from those in the previous chapters unless the same physical meanings are explicitly indicated.

H.-H. Liu, *Fluid Flow in the Subsurface*, Theory and Applications of Transport in Porous Media 28, DOI 10.1007/978-3-319-43449-0_4

4.1 Two Optimality Principles and Their Inconsistency

Optimality principles have been used, as a holistic approach, for flow processes in several important geosystems. Rodriguez-Iturbe et al. (1992) postulated the minimization of energy expenditure rate principles (MEE) at both local and global scales for channel networks in a river basin. At a given location, the energy expenditure rate is represented by water flux multiplied by water energy gradient along the flow direction, as discussed in Chap. 2. The water energy gradient reflects the energy expenditure for a unit amount of water during the water flow process. Without the energy expenditure, the energy gradient should be zero because of the energy conservation. The MEE means that flow structures (e.g., the channel networks) in a system form in such a way that the total energy expenditure rate is minimal across that system.

Rinaldo et al. (1992) developed MEE-based modeling approaches to generate optimal channel networks, and compared their results with those from natural river basins. Striking similarity was observed for natural and optimal networks in their fractal aggregation structures and other relevant features. Using the similar optimality principle, Liu (2011a) developed a group of (partial differential) governing equations for steady-state optimal landscapes (including both channel networks and associated hillslopes) using calculus of variations (Sect. 2.7). Most recently, Liu (2011b) applied the MEE to gravity-dominated unstable (fingering) water flow in unsaturated media (Sect. 2.4). He found that the relative permeability or hydraulic conductivity for unsaturated water flow, in this case, is not only a function of water saturation or capillary pressure [as indicated in the classic theory by Buckingham (1907)], but also a power function of water flux. Furthermore, he showed that conductivities of water flow in both unsaturated soils and on landscapes follow similar power-function relationships (with water flux), but with different exponent values.

On the other hand, the maximum entropy production principle (MEP), initially proposed by Paitridge (1975), has been shown to be useful for predicting behavior of the Earth-atmosphere system. The MEP states that a flow system subject to various flows or gradients will tend towards a steady-state position of the maximum thermodynamic entropy production (Nieven 2010). We refer readers to Glansdorff and Prigogine (1978) for the formulations for entropy production of a flow system involving different physical processes and their mathematical derivations.

Some inconsistency seems to exist between the MEE and the MEP, even they have been successfully applied to different geosystems. Under isothermal conditions, energy expenditure rate is proportional to entropy production rate (Li 2003). Thus, under such conditions, the MEP requires that a river network forms in such a way that the energy expenditure rate for water flow should be at its maximum, which directly contradicts the MEE supported by empirical data for a river basin. This calls for the development of a more precise understanding of fundamental physical laws to reconcile different optimality principles within the context of thermodynamics.

The physical origin of these optimality principles is not totally clear at this point. They cannot be directly deduced from the currently existing thermodynamic laws (Bejan 2000), although some effort has been made to make the connections between the optimality principles and these laws (e.g., Martyushev and Seleznev 2006; Sonnino and Evslin 2007). Bejan (2000) argued that these thermodynamics laws deal largely with processes within systems treated as black-boxes and were not developed for describing internal structures for flow processes in these systems that are associated with the optimality principles. We agree with this assessment. As an effort to develop a general principle to resolve the above issue, Bejan (2000) proposed the constructal law that states that "for a finite-size open system to persist in time (to survive) it must evolve in such a way that it proves easier and easier access to the currents that flow through it". The constructal law has been used to explain many natural phenomena including the related optimality principles (Bejan 2000).

However, a geosystem generally involves a number of flow processes and empirical evidence indicates that not all the flow processes are subject to optimality, as will be discussed below. Liu (2014) proposed a general principle to identify which flow process is subject to optimality for geosystems. It may also help resolve the inconsistence between the MEE and the MEP, the two commonly used optimality principles.

4.2 A Thermodynamic Hypothesis

Liu (2014) proposed the following thermodynamic hypothesis: *a nonlinear natural system that is not isolated and involves positive feedbacks tends to minimize its resistance to the flow process that is imposed by its environment.*

It is important to emphasize that we deal with a system here that is not isolated and subject to mass and/or energy transfer between it and its environment. The system must be nonlinear and involve positive feedbacks such that the relevant flow process plays an important role in forming the flow paths and, at the same time, the formation of these flow paths would further enhance the flow. In such a system, several flow processes (including heat flow or heat transfer) may co-exist. For example, a river basin at least involves surface water flow and soil erosion processes associated with water flow. The hypothesis indicates that the system (river basin) does not tend to provide the minimum resistance to all the flow processes (e.g., soil erosion or soil particle flow), but to the process imposed by its environment (rainfall), or the driving process that is water flow. We will further demonstrate this point in the next section with different geosystems.

The hypothesis is motivated by the notion that natural systems tend to evolve to more and more uniformity (Bejan 2000). A classic example of this is an isolated system in which all the physical properties are completely uniform when it becomes equilibrium. In a system that is not isolated, because of energy and/or mass exchange with its environment, it never becomes completely uniform spatially, but approaches another kind of uniformity that we call "dynamic uniformity" here and may involve emergent structures. Take a river basin as an example. Water entering the river basin has larger energy than the downstream water. Thus, the former has the tendency to equilibrate with the latter. To do so, the former tends to flow to the latter location as quickly as possible and with as little energy loss as possible. Note that the less energy loss means better uniformity (in terms of energy). In this regard, dynamic uniformity corresponds to the MEE. Therefore, the hypothesis is equivalent to: "a nonlinear natural system that is not isolated and involves positive feedbacks tends to maximize the degree of dynamic uniformity for the flow process imposed by its environment." To some extent, the concept of dynamic uniformity is similar to entropy in an isolated system in which entropy tends to be maximized.

The hypothesis also implies the existence of often observed hierarchy structures in some natural systems. This is simply because it applies to both the whole system, such as, for example, a river basin, and a subsystem, such as a part of a river basin associated with a branch of a river network. In this case, the minimum resistance to water flow should occur in the entire river basin and a part of it. As a result, similar flow structures (river/channel geometry) occur at different levels of spatial scales.

The hypothesis specifies the conditions under which optimality occurs for flow processes in geosystems. It is indicated that not all the systems are subject to optimization process except for the nonlinear ones that are not isolated and display positive feedbacks. Especially, for a system involving multiple flow processes, only the driving process (that is imposed by its environment) is subject to resistance minimization.

4.3 Consistence Between the Hypothesis and Flow Behavior in Geosystems

In this section, we demonstrate that the hypothesis is consistent with flow behavior in several typical geosystems (Liu 2014).

4.3.1 Water Flow in Saturated Porous Media

Steady-state saturated flow in porous media, a linear system, is employed here to demonstrate that the minimization of flow resistance, in general, may not occur for a linear system, as implied by our hypothesis.

The energy expenditure rate, *EE*, for water flow in saturated porous media across an entire flow domain Ω can be given by Liu (2011b)

$$EE = \int_{\Omega} K(\nabla H)^2 dV \tag{4.1}$$

where K is hydraulic conductivity, H is hydraulic head, and V is porous-medium volume. Note that the local energy expenditure rate, $K(\nabla H)^2$, is equal to the product of Darcy flux and hydraulic gradient.

By minimizing EE (Eq. 4.1) using Euler-Lagrangian equation (Eq. 2.39), we obtain:

$$\nabla \cdot \nabla H = 0 \tag{4.2}$$

Comparing the above equation with the following water volume (mass) balance equation for water flow in saturated porous media:

$$\nabla \cdot (K \nabla H) = 0 \tag{4.3}$$

one can conclude that the flow process (that always follows Eq. 4.3) satisfies the MEE only for homogeneous porous media, because Eq. 4.2 is the same as Eq. 4.3 for constant K for these media. In fact, this result can also be directly obtained from the well-known Prigogine theorem of the minimization of entropy production which is valid for linear irreversible processes (Prigogine 1955). However, when K is a function of location, which is generally the case for real world problems, steady-state flow processes will not follow the MEE principle, because Eq. 4.2 (or Eq. 4.1) is not identical to Eq. 4.3 anymore. An alternative argument is that for linear systems, there are no feedback mechanisms because, for example, K distributions in saturated porous media are not impacted by water flow at all.

From the above discussion, it is evident that not all flow processes follow optimality principles in terms of flow resistance, especially for linear systems. That is why the hypothesis excludes linear systems.

4.3.2 *Water Flow in Unsaturated Porous Media*

We use unsaturated flow in porous media to show that for a nonlinear system, the existence of positive feedback mechanisms, as indicated by the hypothesis, is required for generating the minimum resistance to water flow.

For a downward, gravity-dominated water flow in unsaturated porous media, fingering (unstable) flow patterns generally occur (Chap. 2). In such a system, the flow process imposed by its environment is water flow because water intrudes into porous media. It is well-known that unsaturated flow is a nonlinear process; the conductivity for water flow is a function of properties associated with water flow itself. Furthermore, positive feedback mechanisms occur here as well. When fingering takes place, more water tends to flow through fingers and therefore makes fingers grow, because fingers provide flow paths with smaller flow resistance than the uniform flow. Thus, the hypothesis indicates that the MEE should apply here.

Based on the MEE, we derived the following relationship for gravity-dominated unsaturated flow in homogeneous porous media in Chap. 2:

$$k_r = F(h)\left(\frac{q}{K_{sat}}\right)^{a_o} \tag{4.4}$$

where k_r is relative permeability for water flow, $F(h)$ is a function of capillary pressure head h, q is the magnitude of Darcy flux, again K_{sat} is saturated hydraulic conductivity, and a_o is a constant whose value seems to be about 0.5 from a limited number of previous studies. The above relationship is supported by experimental observations and is consistent with some empirical models (Liu 2011b). Also note that Eq. 4.4 is identical to Eq. 2.54-2.

A grand challenge facing us in the area of subsurface fluid flow is the need to develop physical laws for large-scale multiphase-flow problems. At a local scale, fluid distribution is mainly controlled by capillarity and is not sensitive to flow conditions. That is why relative permeability at a local scale can be successfully described as a function of saturation (or capillary pressure) only (Buckingham 1907). At a large scale, this is no longer the case, although local-scale relationships have been widely used at large scales because no alternative is available. Equation 4.4 suggests that function forms of large-scale relationships to describe multiphase flow are very likely different from their counterparts at local scales, which cannot be resolved from upscaling based on the same function forms as those at local scales. It is our hope, as discussed in Chap. 2, that the optimality-based approach may provide an important way to obtain such large-scale relationships.

However, the MEE and Eq. 4.4 cannot be applied to nonlinear unsaturated flow in an upward direction, because for such a flow system, there is no positive feedback and therefore no fingering either. Flow structures (such as fingering) are generally a signature of positive feedback mechanisms. Without these mechanisms, small disturbance to water flow that always exists in nature would not grow into flow structures, but rather die out during the flow process. As indicated in the hypothesis, the existence of positive feedback mechanisms is required for the flow resistance to be minimized.

4.3.3 Flow Processes in a River Basin and the Earth-Climate System

Finally, we show that the hypothesis allows for a possible reconciliation between the MEE and the MEP optimality principles, with a focus on the importance of the driving process imposed by the environment (Liu 2014).

The consistency between water flow in a river basin and the hypothesis was already discussed. In this open and nonlinear system, positive feedback exists. During rain, water flows down the hill and tends to flow more in channels/rivers to make them grow once they are initialized by water flow process, because resistance to flow process in these structures is much smaller than that to flow uniformly along a relatively flat hill surface. Also, as previously indicated, water flow is the dominant process imposed by its environment; all other processes involved in a river basin (such as soil erosion) are initialized and driven by water flow. Thus, based on the hypothesis, the MEE applies to water flow in a river basin.

Flow processes in the Earth-atmosphere system are more complex. Many studies indicate that these flow processes follow the MEP (Paltridge 1975; Ozawa et al. 2003). As previously indicated, some inconsistency exists between the MEE and the MEP, despite they have been successfully applied to different geosystems. To clarify this issue, we must recognize that the Earth-atmosphere system generates a minimum flow resistance to (active) "flow process imposed by its environment", but not for other (reactive) processes (e.g., fluid flow) that just facilitates the resistance minimization for the former process, as indicated by the hypothesis. The Earth receives radiation from the hot Sun and transfers a portion of the received heat into space. The atmosphere and oceans act as a fluid system that transports heat from the hot region to the cold one with general circulation (Ozawa et al. 2003), because the convection process is more efficient in transferring heat than the conduction process (Bejan 2000). Obviously, in this system, the "flow process imposed by its environment" is the heat flow; the heat flow process is the initiator of other flow processes.

Under steady-state flow conditions, the average heat flow rate is closely related to entropy production in the Earth-atmosphere system, as shown by Paltridge (1978) and Ozawa et al. (2003), and the MEP corresponds to the maximum convective heat transport (Paltridge 1978). The latter was further confirmed by Clausse et al. (2012) who showed that the temperature distributions on the Earth surface is consistent with a principle (derived from the constructal law) that convective heat flow rate from equator region to the pole region is maximized, or the resistance to heat flow is minimized at a global scale. In this case, the MEP happens to be a by-product of this heat-flow optimization process. Along this line, the MEP in the Earth-atmosphere system and the MEE in a river basin are all consistent with the hypothesis and can be reconciled in terms of minimizing resistance to the "flow process imposed by its environment", or the driving process.

The "maximum heat transfer" hypothesis was initially proposed in a theoretical study of the well-known Rayleigh-Benard convection by Malkus and Veronis (1958). They used that hypothesis to identify wave lengths for the stable convection cells. However, their work has been criticized on the ground that their predicted relationship between the wave length and Rayleigh number is not always consistent with experimental observations (Koschmieder 1993). While the mathematical technique used by Markus and Veronis (1958) is not very reliable for nonlinear problems (Getling 1998), there has been no observation that directly disproves "the maximum heat transfer" hypothesis itself (Koschmieder 1993).

The above discussion has been based on an important argument that the minimum heat-flow resistance results in the maximum heat flow rate and the MEP in the Earth-atmosphere system. This argument can be further justified as follows. Under steady-state flow conditions, the entropy production in the Earth-atmosphere system is given by Paltridge (1978) and Ozawa et al. (2003):

$$S' = \int_{A*} \frac{dQ}{T} = Q\left(\frac{1}{T_L} - \frac{1}{T_H}\right) = Q\frac{\Delta T}{T_{av}^2\left[1 - \left(\frac{\Delta T}{2T_{av}}\right)^2\right]} \tag{4.5}$$

where S' is the entropy production rate, $A*$ is the boundary (Earth surface), Q is heat flow through the boundary, T_L and T_H are average Earth surface temperatures in the low and high temperature regions, respectively, T_{av} is the average temperature of both high and low temperature regions, and $\Delta T = T_H - T_L$. (Heat flows from the high-temperature region to the low temperature region.) The above equation represents the fact that the entropy production by some processes associated with turbulence is completely discharged into the surrounding system through the boundary under the steady-state condition (Ozawa et al. 2003).

Since $[\Delta T/(2T_{av})]^2$ is generally on the order of 1 % and T_{av} can be reasonably determined based on global solar heat current and Earth surface radiation into space only (Clausse et al. 2012), Eq. 4.5 can be rewritten as

$$S' \approx Q\frac{\Delta T}{T_{av}^2} \tag{4.6}$$

where T_{av} is considered a well constrained parameter that does not depend on Q and ΔT. Clausse et al. (2012) indicated that convective flow rate from the high temperature region to thc low temperature region can be expressed as

$$Q = C(\Delta T)^{3/2} \tag{4.7}$$

where C is a constant related to Earth radius and fluid properties involved in heat flow. Accordingly, the conductance for heat flow will be:

$$K_H = \frac{Q}{\Delta T} = C(\Delta T)^{1/2} \tag{4.8}$$

Based on Eqs. 4.6, 4.7 and 4.8, we have

$$S' = \frac{Q^{5/3}}{T_{av}^2 C^{2/3}} \tag{4.9}$$

$$K_H = C^{2/3}Q^{1/3} \tag{4.10}$$

Clearly, Eqs. 4.9 and 4.10 show that the maximum K_H (or the minimum heat flow resistance) corresponds to the maximum heat flow rate (Q) and entropy production rate (S'). It is also of interest to note that the formulation for the heat conductance (Eq. 4.10) is consistent with the conductivity relations for water flow in both river basins and unsaturated porous media, derived from the optimality principle (Chap. 2), because they all are power functions of flux (or flow rate) with positive exponent values.

4.3.4 A Further Discussion on the MEP

The thermodynamic basis for MEP has been a subject of active research in the literature (Paltridge 2009; Dewar 2003, 2005; Martyushev and Seleznev 2006). As previously indicated, the MEP is not a fundamental physical principle, because it is not consistent with the MEE. A fundamental physical principle should be able to explain observations from different areas. Our hypothesis is able to reconcile MEP and MEE. The discussion on the MEP in Sect. 4.3.3 is based on the result of Clausse et al. (2012), or Eq. 4.7 that was derived from the physics of fluid circulation between the low temperature and high temperature regions on the Earth surface and thus is appropriate for the discussion there. Given that considerable attention has been given to seeking the thermodynamic basis of the MEP in the scientific community, we further demonstrate in this subsection that our hypothesis allows for deriving the MEP even under a more general case than that defined by Eq. 4.7.

There are two heat transfer mechanisms between the low- and high-temperature regions on the Earth's surface: conduction and convection. The former is relatively small compared with the convection and therefore ignored here for the simplicity (e.g., Clausse et al. 2012). (Inclusion of the conduction will not alter the final conclusion.) Under the steady-state fluid flow condition, fluid with temperature T_H from the high-temperature region flows to the low-temperature region. At the same time, the same amount of fluid with temperature T_L flows from the low-temperature region to the high-temperature region. As a result, the net heat flow rate between the two regions is given by

$$Q \propto Q_{fl} \Delta T \tag{4.11}$$

where Q_{fl} is the fluid flow rate between the two regions.

By definition (Eq. 4.8), the heat conductivity between the two regions can be obtained from Eq. 4.11:

$$K_H \propto Q_{fl} \tag{4.12}$$

Note that fluid flow between the low- and high-temperature regions is induced by the temperature difference $\Delta T = T_H - T_L$, or

$$Q_{fl} = f_Q(\Delta T) \tag{4.13}$$

The temperature difference results in fluid density differences, and consequently fluid pressure differences (through gravity) between the two regions. The latter gives rise to the fluid flow. At this point, we do not need to know details of function f_Q, such as that given in Eq. 4.7. It is adequate to know that Q_{fl} is a monotone increasing function of the temperature difference. (Figure 4.1 shows an example of

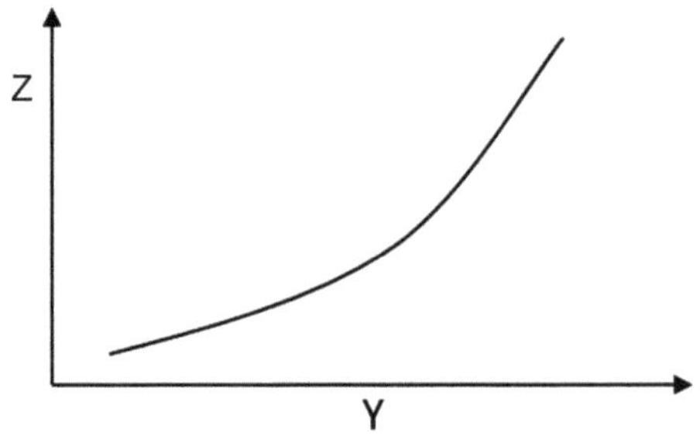

Fig. 4.1 Z is a monotone increasing function of Y. Obviously Y is a monotone increasing function of Z as well

monotone increasing function.) It is easy to understand physically that a larger temperature difference corresponds to a larger fluid flow rate between the low- and high-temperature regions. In fact, it is how the positive feedback comes to play in the system.

Equations 4.5 and 4.11–4.13 yield the entropy production as

$$S' \propto Q\Delta T \propto Q_{fl}\Delta T^2 = f_Q(\Delta T)\Delta T^2 = F(\Delta T) \tag{4.14}$$

Clearly, function F is also a monotone increasing function of the temperature difference because f_Q is a monotone increasing function. From Eq. 4.14, we have

$$\Delta T = F^{inv}(S') \tag{4.15}$$

where superscript *inv* refers to inverse function. The function F^{inv} is the inverse function of F. Again, F^{inv} should be a monotone increasing function when F is a monotone increasing function, as demonstrated in Fig. 4.1.

A combination of Eqs. 4.12, 4.13 and 4.15 results in

$$Q_{fl} = f_Q(F^{inv}(S')) = G(S') \tag{4.16}$$

and

$$K_H \propto G(S') \tag{4.17}$$

where function G is obviously a monotone increasing function, because both f_Q and F^{inv} are monotone increasing functions. By the nature of a monotone increasing function, S' and K_H should reach their maximum values at the same time. This again supports the notion that the MEP in the Earth-atmosphere system is not a fundamental physical principle, but a by-product of maximizing K_H (or minimizing resistance to heat flow between the low- and high-temperature regions).

4.4 Consistency Between the Hypothesis and Darwin's Evolution Theory

It is of interest to indicate that the proposed hypothesis is consistent with Darwin's evolution theory, although the latter is for genetic phenomena. At a given time, biological creatures always exhibit some physical and behavioral variations on different levels (e.g., individuals in a community, organisms, DNA and genes). These variations result from random mutations that are changes in the genetic sequence. Mutations are either harmful, neutral, or in some cases beneficial to the organism. Through the process of natural selection, beneficial mutations will be selected for and harmful ones will be selected against as the affected genes move through the subsequent generations.

Table 4.1 shows that similarity exists between some key elements of Darwin's evolution theory and the proposed hypothesis. Natural perturbation in fluid and medium properties, although not explicitly mentioned in our hypothesis, is well known to exist for all geological flow systems. This perturbation may be small in many cases, but serves as the seed to generate flow patterns with the minimum flow resistance. Thus, "natural perturbation" corresponds to "mutation" in Darwin's theory.

For a geological flow system, if a perturbation results in a slight change in the corresponding flow pattern that reduces the flow resistance, this slight change will be "remembered" such that it is enhanced in the next moment to further reduce the resistance, which is a result of positive feedback. This is how fingering in unsaturated flow systems and flow channels in a river basin are formed. (Note that a negative feedback tends to reduce the perturbations with time.) Thus, the positive feedback mechanism is very similar to the procedure for the next generation to inherit the beneficial mutations. Finally, the process to achieve the minimum flow resistance in a geological system is obviously comparable to the natural selection process to make the next generation to be better adapted to its environment.

It is appropriate to state that some nonlinear flow processes in geological systems are "alive" and very well evolved. We need to respect them to live in harmony with them in the world which we share with them. We have to learn their "language" first to predict their behavior using it. Otherwise, our model prediction of natural processes will not be very meaningful. We believe that optimality is an important part of their "language". Discussions in this chapter and Chap. 2 are examples of efforts to learn and make use of the "language" of the relevant nonlinear flow processes.

Table 4.1 Similarity of some key elements of Darwin's evolution theory and the proposed hypothesis

Darwin's evolution theory	The proposed hypothesis
The mutation	The natural perturbation
The inheriting beneficial mutations	The positive feedback
The natural selection	The minimum flow resistance

4.5 Calculation of Inelastic Deformation of Natural Rock

A fundamental aspect of our hypothesis is to first identify the driving process, because a system tends to evolve minimum resistance only to the driving process imposed by its environment. While the hypothesis was initially proposed for fluid flow in geosystems (Liu 2014), it has much broader applications. To some extent, it is related to Darwin's evolution theory, as previously discussed. In this section, we present a new approach to calculate the inelastic deformation of rocks based on the hypothesis.

When a rock is under stress, its deformation is initially relatively small and elastic. Chapter 3 provides a detailed study on the elastic deformation with the two-part Hooke model (TPHM). With increasing stress, the deformation becomes inelastic and not reversible. There are two major mechanisms for inelastic deformation: plasticity and damage. Plasticity refers to the dislocation or "slip" of the solid materials at microscale and damage refers to initialization and propagation of micro-cracks. The former is generally more important for metals and the latter for natural rock. Nevertheless, the inelastic deformation of natural rock is a combination of plastic deformation and damage. In this section, we do not distinguish between them.

There are a number of ways to determine inelastic deformation of natural rock (Krajcinovic 1989). We investigate this issue here within the context of thermodynamics. The first step is to determine the driving process for the inelastic deformation that is associated with the driving force. There are different processes occurring in inelastic deformation, including change in strain, small-scale dislocation, initialization and propagation of micro-cracks, and creation of new surfaces within the rock. The driving force is obviously due to the stress and the result is the stress-driven change in inelastic strain. Let us suppose that the "purpose" of the inelastic strain development is to minimize the resistance to further changes in strain.

For inelastic deformation, our hypothesis is mathematically equivalent to minimizing the following entropy production (*EP*) for a given change in strain:

$$EP = \sum_{i=1}^{3} \sigma_i d\varepsilon_i^{ine} \tag{4.18}$$

where σ_i and ε_i^{ine} are principal stress and inelastic strain in the i directions, respectively. It is common to determine the relationship between stress and inelastic strain in stress space (e.g., Hansen and Schreyer 1994). In other words, in Eqs. 4.18 and 4.22 to be discussed later, $d\varepsilon_i^{ine}$ and ε_i^{ine} are considered to be fixed. It is important to note that the *EP* actually reflects the resistance to inelastic deformation. For a given $d\varepsilon_i^{ine}$, a smaller *EP* gives smaller σ_i (or the driving force). On the other hand, the expression of *EP* (Eq. 4.18) does not include the temperature-related coefficients that exist in the expression of the entropy production given by Glansdorff and Prigogine (1978). This treatment is just for convenience, because we assume the rock system to be isothermal in this section.

The MEP has been used previously to determine inelastic deformation (Ziegler 1963; Hansen and Schreyer 1994; Collins and Houlsby 1997). In these conventional MEP applications, entropy production includes contributions from all the involved processes. As previously indicated, the MEP is not a fundamental principle governing a flow process; inelastic deformation is a "flow process". Equation 4.18 deals with the driving process only, because the optimality should not be applied to the other processes.

The minimization of *EP* (Eq. 4.18) is subject to a number of constraints such that the minimization results are physically valid and consistent with the corresponding experimental observations. Because the inelastic deformation occurs on a yield surface in stress space, the minimization result should satisfy the following equation for the yield surface that is experimentally determined:

$$f_y(\sigma_1, \sigma_2, \sigma_3, \kappa) = 0 \tag{4.19}$$

where κ is a hardening parameter that is generally an empirical function of inelastic strain (Fjær et al. 2008).

Other constraints are the related experimental observations except the yield surface. These observations can be expressed as

$$F_j(\sigma_1, \sigma_2, \sigma_3, \varepsilon_1^{ine}, \varepsilon_2^{ine}, \varepsilon_3^{ine}) = 0 \quad (j = 1, 2 \ldots N) \tag{4.20}$$

where N is the total number of equations describing the observations. One of the important observed relations for natural rock is the inelastic volumetric strain, because it is closely related to propagation of fractures and change in their apertures. Consequently, it has an important impact on fluid flow through the rock. The volumetric strain (ε_v^{ine}) can be expressed as a function (F_v) of related observations

$$\varepsilon_v^{ine} - F_v(\sigma_1, \sigma_2, \sigma_3, \varepsilon_1^{ine}, \varepsilon_2^{ine}, \varepsilon_3^{ine}) = 0 \tag{4.21}$$

Based on Eqs. 4.18–4.20, the corresponding Lagrangian is given as

$$L = \sum_{i=1}^{3} \sigma_i d\varepsilon_i^{ine} + \lambda_0 f_y + \sum_{j=1}^{N} \lambda_j F_j \tag{4.22}$$

where λ_0 and λ_j are Lagrangian multipliers. From

$$\frac{\partial L}{\partial \sigma_i} = 0 \tag{4.23}$$

one can obtain:

$$d\varepsilon_i^{ine} = -\lambda_0 \frac{\partial f_y}{\partial \sigma_i} - \sum_{j=1}^{N} \lambda_j \frac{\partial F_j}{\partial \sigma_i} \tag{4.24}$$

Lagrangian multipliers can be obtained by solving combined Eqs. 4.19, 4.20 and 4.24. These equations provide a general way to calculate inelastic strain based on our hypothesis. The calculation results satisfy all the experimental observations through Eqs. 4.19 and 4.20. Mathematically, it is important that Eq. 4.20 is given in the stress space, or it includes stresses in its expression. Otherwise, the constraints would not be able to be imposed, as implied by the last term on the right hand side of Eq. 4.24.

Traditionally, the inelastic strain is often calculated by

$$d\varepsilon_i^{ine} = -\lambda_0 \frac{\partial \Phi}{\partial \sigma_i} \tag{4.25}$$

where Φ is a potential function. Comparison between Eqs. 4.24 and 4.25 leads to

$$\lambda_0 d\Phi = \lambda_0 df_y + \sum_{j=1}^{N} \lambda_j dF_j \tag{4.26}$$

When $\Phi = f_y$, the inelastic deformation is said to follow the "associated flow" rule. Otherwise, it would follow "non-associated flow" rule. Equation 4.26 clearly explains why the associated flow rule, in general, is not the case for natural rocks because it is not able to incorporate all the experimental observations (Eq. 4.20) except the yield surface (Eq. 4.19).

The traditional approach to choose potential function Φ is largely through a trial and error process. Mathematical form of Φ and values of the related parameters in the function are empirically adjusted to match the corresponding experimental observations. Our method deals with the issue from a very different angle that is based on the proposed thermodynamic hypothesis. We start directly with experimental observations with which we construct functions f_y and F_j. Then without trial and error, Φ can be straightforwardly obtained from Eq. 4.26 (together with Eqs. 4.19, 4.20 and 4.24). Thus, the procedure to calculate the inelastic deformation, outlined in this section, is more straightforward.

4.6 Concluding Remarks

(A) This chapter presents a thermodynamic hypothesis regarding optimality principles for flow process in geosystems. It states that *a nonlinear natural system that is not isolated and involves positive feedbacks tends to minimize its resistance to the flow process that is imposed by its environment.*

(B) Consistence between the hypothesis and typical flow processes in geosystems is demonstrated. In spirit, the hypothesis is consistent with Darwin's evolution theory. The hypothesis reconciles the seeming inconsistence between the MEE and the MEP. An application of the hypothesis to calculation of inelastic deformation of natural rock is also outlined.

(C) The hypothesis is fundamental in nature, but proposed in a phenomenological manner. Further examinations of the usefulness and potential limitations of the hypothesis in describing other processes, distinct from flow processes in geosystems, are needed.

References

Bejan A (2000) Shape and structure: from engineering to nature. Cambridge University Press, New York

Buckingham E (1907) Studies on the movement of soil moisture. Bulletin 38, USDA Bureau of Soils, Washington DC

Clausse M, Meunier F, Reis AH et al (2012) Climate change: in the framework of the constructal law. Int J Global Warming 4:242–260

Collins IF, Houlsby GT (1997) Application of themomechanical principles to the modeling of geotechnical materials. Proc R Soc London, Ser A, Math Phys Eng Sci 453(1964):1975–2001

Dewar RC (2003) Information theory explanation of the fluctuation theorem, maximum entropy production, and self organized criticality in non-equilibrium stationary states. J Phys A: Math Gen 36:631–641

Dewar RC (2005) Maximum entropy production and the fluctuation theorem. J Phys A: Math Gen 38:371–381

Eigen M (2013) Strange simplicity to complex familiarity. Oxford University Press, Oxford, UK

Fjær E, Holt RM, Horsrud P (2008) Petroleum related rock mechanics, 2nd edn. Elsevier, Boston

Getling AV (1998) Rayleigh Benard convection: structures and dynamics. World Scientific, Singapore

Glansdorff P, Prigogine I (1978) Thermodynamic theory of structure, stability and fluctuations. Wiley-Interscience, New York

Hansen NR, Schreyer HL (1994) A thermodynamically consistent framework for theories of elastoplasticity coupled with damage. Int J Solids Structures 31(3):359–389

Koschmieder KL (1993) Benard cells and Taylor vortices. Cambridge University Press, New York

Krajcinovic D (1989) Damage mechanics. Mech Mater 8:117–197

Li GA (2003) Study on the application of irreversible thermodynamics in the problems of porous medium seepage. Dissertation for Doctoral Degree, Zhejiang University, China

Liu HH (2011a) A note on equations for steady-state optimal landscapes. Geophys Res Letter. doi:10.1029/2011GL047619

Liu HH (2011b) A conductivity relationship for steady-state unsaturated flow processes under optimal flow conditions. Vadose Zone J. doi:10.2136/vzj2010.0118

Liu HH (2014) A thermodynamic hypothesis regarding optimality principles for flow processes in geosystems. Chin Sci Bull 59(16):1880–1884

Malkus WVR, Veronis G (1958) Finite amplitude cellular convection. J Fluid Mech 4:225–260

Martyushev LM, Seleznev VD (2006) Maximum entropy production principle in physics, chemistry and biology. Phys Rep 426:1–45

Nieven RK (2010) Minimization of a free-energy-like potential for non-equilibrium flow systems at steady state. Phil Trans R Soc B2010(365):1213–1331. doi:10.1098/rstb.2009.0296

Ozawa H, Ohmura A, Lorenz R et al (2003) The second law of thermodynamics and the global climate system: a review of the maximum entropy production principle. Rev Geophys. doi:10.1029/2002RG000113

Paltridge GW (1975) Global dynamics and climate—a system of minimum entropy exchange. Quart J R Met Soc 101:475–484

Paltridge GW (1978) The steady-state format of global climate. Quart J R Met Soc 104:927–945

Paltridge GW (2009) A Story and a recommendation about the principle of maximum entropy production. Entropy. doi:10.3390/e11040945

Prigogine I (1955) Introduction to thermodynamics of irreversible processes. Interscience Publishers, New York

Rinaldo A, Rodriguez-Iturbe I, Rigon A et al (1992) Minimum energy and fractal structures of drainage networks. Water Resour Res 28:2183–2191

Rodriguez-Iturbe I, Rinaldo A, Rigon A (1992) Energy dissipation, runoff production and the three-dimensional structure of river basins. Water Resour Res 28(4):1095–1103

Sonnino G, Evslin J (2007) The minimum rate of dissipation principle. Phys Lett A 365:364–369

Ziegler H (1963) Some extremum principles in irreversible thermodynamics with application to continuum mechanics. In: Sneddon IN, Hill R (eds) Progress in solid mechanics, vol 4. Interscience Publishers, New York

Chapter 5
Final Remarks: An "Unfinished" Book

Abstract This chapter summarizes key results from previous chapters. While these chapters document the important progresses in generalizing several physical laws related to subsurface fluid flow and mechanical deformation, they may or may not provide the final forms of the generalizations. Thus, this book should be viewed as a starting point for revisiting these physical laws and for stimulating further research activities in the related areas.

This book deals with the physical laws related to subsurface fluid flow processes. The major results from the previous chapters are recapped as follows.

Darcy's law is the fundamental law for subsurface fluid flow. However, for low-permeability media, Darcy's law does not always hold because of the strong solid-liquid interaction. Chapter 1 presents a generalized Darcy's law given as

$$q = K\left[i - \frac{I}{\gamma(\frac{1}{\alpha})}\gamma\left(\frac{1}{\alpha}, \left(\frac{i}{I^*}\right)^{\alpha}\right)\right] \tag{1.21-1}$$

where the threshold hydraulic gradient I and parameter α characterize impacts of solid-liquid interaction and pore size distribution, respectively, on liquid flow. The above relationship is very general in a sense that it includes Darcy's law and the two commonly used forms of modified Darcy's law as its special cases. The consistency between it and experimental observations from different sources is demonstrated.

The Darcy-Buckingham law was developed based on the local-equilibrium condition that, however, does not always hold especially when fingering flow occurs. Chapter 2 is devoted to generalizing this important law by relaxing the local-equilibrium condition. The new development is based on an optimality principle that water flow in unsaturated media self-organizes in such a way that the resistance to water flow is minimized. The key result is given as

$$k_r(h, S_*) = F(h)\left(\frac{q}{K}\right)^{a_O} \tag{2.54-2}$$

H.-H. Liu, *Fluid Flow in the Subsurface*, Theory and Applications of Transport in Porous Media 28, DOI 10.1007/978-3-319-43449-0_5

Unlike the traditional form of the Darcy-Buckingham law, the relative permeability in the above equation is also a function of water flux. That is a direct result of the non-equilibrium flow behavior. The generalized Darcy-Buckingham law has been shown to be consistent with laboratory observations and field data for gravitational fingering flow in unsaturated soils. This generalization is the theoretical foundation for the active fracture model, the key constitutive relationship for modeling flow and transport in the unsaturated zone of Yucca Mountain, Nevada, that is the national geological disposal site for high-level nuclear waste in USA.

Hooke's Law is commonly used for modeling elastic deformation of solids. However, natural rock has unique features compared with other solids one of which is the existence of small-scale deformation heterogeneities (such as micro-cracks). To address this issue, Chap. 3 presents the TPHM given by Eq. 3.8 and its variations:

$$-\frac{dV}{V_0} = \gamma_e \frac{d\sigma}{K_e} + \gamma_t \exp\left(-\frac{\sigma}{K_t}\right) \frac{d\sigma}{K_t} \tag{3.8}$$

Without the soft part, the TPHM is reduced to the conventional form of Hooke's Law. The development is based on a well-known fact that micro-cracks (soft part) exist in natural rock and exhibit very different mechanical behavior from the rest of rock mass. Note that micro-cracks can serve as the dominant flow pathways within a rock matrix.

Optimality principles have been used in many different areas, including the generalization of Darcy-Buckingham law documented in this book. As discussed in Chap. 2, these principles have a long history in science and have been proven to be useful for many engineering applications. It is, however, surprising that the thermodynamic base for these principles has not been fully established. Chapter 4 proposes a thermodynamic hypothesis regarding optimality principles for flow process in geosystems. It states that *a nonlinear natural system that is not isolated and involves positive feedbacks tends to minimize its resistance to the flow process that is imposed by its environment.* Consistency between the hypothesis and typical flow processes in geosystems is demonstrated. In spirit, the hypothesis is also consistent with Darwin's evolution theory.

However, as indicated by the title of this chapter, this book is not a really "finished" book in many ways. While the book documents the important progresses in generalizing several physical laws related to subsurface fluid flow and mechanical deformation, it may or may not provide the final forms of the generalizations. Thus, this book should be viewed as a starting point for revisiting these laws and for stimulating further research activities along the line. It is the author's hope that the "book" will be eventually "finished" by future scientists or engineers, likely including readers of this book, who have better physical insights and out-of-box ideas.

Index

H.-H. Liu, *Fluid Flow in the Subsurface*, Theory and Applications of Transport in Porous Media 28, DOI 10.1007/978-3-319-43449-0

MIX
Papier aus verantwortungsvollen Quellen
Paper from responsible sources
FSC® C105338

If you have any concerns about our products,
you can contact us on
ProductSafety@springernature.com

In case Publisher is established outside the EU,
the EU authorized representative is:
Springer Nature Customer Service Center GmbH
Europaplatz 3, 69115 Heidelberg, Germany

Printed by Libri Plureos GmbH
in Hamburg, Germany